The
Pollination of Flowers
by
Insects

Reports of Linnean Symposia

Speciation in Tropical Environments
(Lowe-McConnell)

Biological Journal of the Linnean Society
Vol. *1* 1969 pp. 1–246

New Research in Plant Anatomy
(Robson, Cutler & Gregory)

Supplement 1 to the *Botanical Journal of the Linnean Society* Vol. *63* 1970

Early Mammals
(Kermack & Kermack)

Supplement 1 to the *Zoological Journal of the Linnean Society* Vol. *50* 1971

The Biology and Chemistry of the Umbelliferae (Heywood)

Supplement 1 to the *Botanical Journal of the Linnean Society* Vol. *64* 1971

Behavioural Aspects of Parasite Transmission (Canning & Wright)

Supplement 1 to the *Zoological Journal of the Linnean Society* Vol. *51* 1972

The Phylogeny and Classification of the Ferns (Jermy, Crabbe & Thomas)

Supplement 1 to the *Botanical Journal of the Linnean Society* Vol. *67* 1973

Interrelationships of Fishes
(Greenwood, Miles & Patterson)

Supplement 1 to the *Zoological Journal of the Linnean Society* Vol. *53* 1973

The Biology of the Male Gamete
(Duckett & Racey)

Supplement 1 to the *Biological Journal of the Linnean Society* Vol. *7* 1975

Continued as the Linnean Society Symposium Series

No. 1

The Evolutionary Significance of the Exine (Ferguson & Muller) (1976)

No. 2

Tropical Trees. Variation, Breeding and Conservation (Burley & Styles) (1976)

No. 3

Morphology and Biology of Reptiles (Bellairs & Cox) (1976)

No. 4

Problems in Vertebrate Evolution (Andrews, Miles & Walker) (1977)

No. 5

Ecological Effects of Pesticides (Perring & Mellanby) (1977)

No. 6

this volume (1978)

Also published

Botanical Journal of the Linnean Society, Vol. *73*, Nos 1–3 July/Sept./Oct. 1976
Zoological Journal of the Linnean Society, Vol. *63*, Nos 1 & 2 May/June 1978

The Biology of Bracken (Perring & Gardiner)

Sea Spiders (Pycnogonida) (Fry)

(*Facing title page*)

FRONTISPIECE
Floral colours as humans see them (right) and as insects (left) might see them.
(For full explanation see Plate 1, Kevan, p. 78.)

Linnean Society Symposium Series

Number 6

The Pollination of Flowers by Insects

Editor

A. J. Richards

Department of Plant Biology,
University of Newcastle upon Tyne

Published for the Linnean Society of London by Academic Press

ACADEMIC PRESS INC. (LONDON) LIMITED
24/28 Oval Road
London NW1 7DX
(Registered Office)
(Registered number 5985 14)

US edition published by
ACADEMIC PRESS INC.
111 Fifth Avenue
New York
New York 10003

ISBN 0–12–587460–X
LCCCN 77–93488

Printed in Great Britain by
Henry Ling Ltd., The Dorset Press, Dorchester, Dorset

Foreword

This volume is the 16th report of a symposium which the Linnean Society has produced in the past nine years. It prints the papers given at a meeting organised in association with the Botanical Society of the British Isles at Newcastle upon Tyne in 1977.

Its appearance is welcome for several reasons. For one thing, the Linnean is concerned with the whole gamut of biology, the forum for the common ground of ever-increasing specialisations; and here are zoologists and botanists exploiting their techniques and knowledge to their mutual enrichment.

Nor is it only the co-operation of the varied scientists from home and abroad who took part that is so promising: it is the co-operation also of these two eminent Societies which, not for the first time, produced such happy effects. The Botanical Society of the British Isles, the largest of its kind in our islands, has a preponderance of amateurs among its members, which means a notable extension of interest in the subject. The Linnean provided more professional backing and the prestige of its 190 years of fertile activity, the oldest biological society in the world.

To this, I would like to add my special personal pleasure in this particular joint venture. This is not only because I much enjoyed attending the conference, but because I have been a member of the Botanical Society for 43 years, and was its President for the usual two; and a Fellow of the Linnean for 25 years, and one of its Vice-Presidents for an unusual consecutive three. So my heart is very much with both.

John Richards has been a model organiser and editor, and merits the warm thanks of us all. The success of the meeting, was largely due to him; and the ease with which this volume has been prepared is due to him too, and to the unvarying quiet efficiency of Michael Ewins at Academic Press.

I trust that this volume will justify the enthusiasm with which Academic Press, with their long experience, agreed to publish it and that it will be as widely appreciated and valued as it deserves.

DAVID McCLINTOCK
Editorial Secretary
Linnean Society of London

Contributors

BEATTIE, A. *Department of Biological Sciences, Northwestern University, Evanston, Illinois 60201, U.S.A.* (p. 151)

BRANTJES, N. B. M. *Botanisches Laboratorium, Catholic University of Nijmegen, Toernooiveld, Nijmegen, The Netherlands* (p. 13)

CORBET, SARAH A. *Department of Applied Biology, Pembroke Street, Cambridge, U.K.* (p. 21)

EISIKOWITCH, D. *Department of Botany, University of Tel-Aviv, Tel-Aviv, Israel* (p. 125)

FAEGRI, K. *P.O. Box 12, N-5014 Bergen, Norway* (p. 5)

FAULKNER, G. J. *National Vegetable Research Station, Wellesbourne, Warwick CV35 9EF, U.K.* (p. 201)

IBRAHIM, HALIJAH *Department of Plant Biology, Ridley Building, University of Newcastle upon Tyne, Newcastle upon Tyne NE1 7RU, U.K.* (p. 165)

KAY, Q. O. N. *Department of Botany and Microbiology, University College, Swansea, South Wales SA2 8PP, U.K.* (p. 175)

KEVAN, P. G. *Department of Biology, University of Colorado, Colorado Springs, Colorado 80907, U.S.A.* (p. 51)

LEVIN, D. A. *Department of Botany, University of Texas, Austin, Texas 78712, U.S.A.* (p. 133)

MEEUSE, A. D. J. *Hugo de Vries Laboratorium, Plantage Middenlaam 2a, (NL) 1018 DD, Amsterdam, The Netherlands* (p. 47)

MEEUSE, B. J. D. *Botany Department, University of Washington, Seattle, Washington 98122, U.S.A.* (p. 97)

MOGFORD, D. J. *Department of Plant Sciences, Rhodes University, P.O. Box 94, Grahamstown 6140, South Africa* (p. 191)

PIJL, L. VAN DER *236 Sportlaan, The Hague, The Netherlands* (p. 79)

PROCTOR, M. C. F. *Department of Biological Sciences, University of Exeter, Hatherly Laboratories, Prince of Wales Road, Exeter EX4 4PS* (p. 105)

RICHARDS, A. J. *Department of Plant Biology, Ridley Building, University of Newcastle upon Tyne, Newcastle upon Tyne NE1 7RU, U.K.* (pp. 1, 165)

STELLEMAN, P. *Hugo de Vries Laboratorium, Plantage Middenlaam 2a, (NL) 1018 DD, Amsterdam, The Netherlands* (p. 41)

VALENTINE, D. H. *Department of Botany, The University, Manchester M13 9PL, U.K.* (p. 117)

VOGEL, ST. *Botanisches Institut der Universität, Rennweg 14, A-1030 Vienna, Austria* (p. 89)

WOODELL, STANLEY R. J. *Botany School, University of Oxford, Oxford, U.K.* (p. 31)

Preface

In 1973 the Department of Plant Biology at the University of Newcastle upon Tyne moved into a fine new building. I thought it would be nice to celebrate this with a major conference, and suggested the subject of plant–insect interactions, a forward-looking field of current interest in which I myself was involved, to the Botanical Society of the British Isles (BSBI).

The idea was welcomed with enthusiasm, but in view of other conferences, it was restricted to the pollination of flowers by insects. This has recently taken on a more modern lease of life, stimulated by two excellent books, *The Principles of Pollination Ecology* by K. Faegri and L. van der Pijl, and *The Pollination of Flowers* by M. C. F. Proctor and P. F. Yeo. The first three authors addressed the conference, and Dr Yeo would have been present, but was unfortunately abroad.

Many other leading workers in the field were asked to attend and give papers, and nearly all agreed to do so. In all, some 150 participants (50 from abroad) assembled during cold and unseasonable weather at Castle Leazes Hall in the University from 11th to 14th April 1977. Eighteen papers were read, and these form the substance of this volume. In addition, a number of films were shown, and there were many exhibits. The conference was indeed both enjoyable and instructive in every way, and I would like to thank all those who attended for making it such a success. It was funded in part by the Royal Society and by the British Council, to both of whom I again express my thanks for their support. Unfortunately it did not prove financially possible to bring Dr D. Levin from Texas. But the paper he had prepared was read and is published here.

The Linnean Society joined the Botanical Society in organising the conference, primarily to add their experience and expertise in publishing the Proceedings; and I should like to thank Mr David McClintock, their Editorial Secretary, in particular, for his help. I should also mention Dr Michael Proctor, who, in addition to the paper he gave, presented a characteristically magnificent series of his remarkable photographs to illustrate this Volume.

This report derives from the conference, and I must thank Mr E. M. Caldwell, Miss A. M. Biggins, Miss G. Williams and other members of the Department, for much hard work before, during and after it. Finally, Mrs Mary Briggs, the Honorary General Secretary of the Botanical Society of the British Isles was, as always, a tower of strength throughout.

A. J. RICHARDS
University of Newcastle upon Tyne

Contents

xi

Introduction

A. J. RICHARDS

Department of Plant Biology,
University of Newcastle upon Tyne

The study of the pollination biology of plants is a discipline of long standing. It is now nearly two hundred years since Sprengel published the classic first systematic account of adaptions of flowers to pollination (*Das entdeckte Geheimniss der Natur im Bau und in der Befruchtung der Blumen* (1793)). Later, but still more than a century ago, Darwin studied such relationships in great detail and in a wide range of plants and insects. His observations on orchids and Primulas are best known, and in each case formed the basis of considerable research into situations of great fascination and complexity during the following one hundred years. In these and in much of his other work on pollination biology, Darwin typically displayed a modernity of approach, a thoroughness in establishing facts, and penetration in understanding complex problems. His writings on pollination biology include some of the best work published in the subject right up to the present day.

Despite an early start, the subject seemed latterly to lose its initial impetus. Exhaustive works by Müller, Knuth and others provided a good deal of the initial descriptive framework of the subject, together with a substantial ecological and evolutionary insight at a time when ecology as a discipline had scarcely been conceived. Much of the early work was published in the German language, and this may partly account for the relative lack of interest in the subject which has persisted in Anglo-American biology for the last hundred years. More recently, excellent books by Faegri and van der Pijl in several editions (with yet another promised soon) and by Proctor and Yeo have provided up to date and comprehensive summaries of the subject which have done much to reawaken interest in the English speaking world. Indeed, it would have been possible to restrict the present conference to speakers of British origin, and still maintain a quality and interest: it is doubtful whether this could have been so twenty, or even ten years ago.

However, with notable exceptions, such as the work of Bateman, and of Free, the subject remained primarily descriptive, and had scarcely entered the secondary, experimental phase of a science ten years ago. Perhaps this was due to the intrinsic interest of many phenomena in pollination biology, including some of the most charismatic stories in the whole of biology, and due also to the complexity and variability of these phenomena, making the formulation of generalisations and "laws" as difficult as in any other field of ecology.

Particularly noticeable was the self-centredness of the subject, more character-istic of a physiological than an organismal discipline. Most workers seemed to consider pollination relationships as ends in themselves, without regard to their potential in understanding other phenomena (insect behaviour and nerve physio-logy); in providing data to fit more general models (predator guilds and niche specificity); in providing phenomena suitable for the study of the adaptiveness of physiological phenomena (aroid sapromyophily); or as a tool for the study of population biology, ecological genetics and evolution. Only in the last ten years or so have these, and other fields developed, in some cases to a spectacular extent. Some of these developments, and the accompanying establishment of sophisticated experimental techniques now employed by floral biologists are described by Faegri in the first introductory paper in this volume, and some are elaborated upon in later papers. It is my hope that this volume will help to show lingering doubters in other disciplines that pollination biology has not only rapidly caught up with them in concept and techniques, but offers fascinating, and very con-siderable potential for workers in other fields. It has always been among the most exciting and intrinsically interesting of subjects. Now there is a real chance that it may become among the most important as well. Biology has passed through its descriptive and experimental phases. But unlike other sciences, its subjects do not act in isolation, but in a structured multidimensional environment consisting importantly of other organisms. A last great challenge, and a potential solution to basic problems, including famine, disease and species extinction, is the applica-tion of modern scientific understanding to inter-organism interactions. Pollination biology provides a wealth of examples of such interactions, from the highly evolved (and often simple), to the generalistic (and complex), exhibiting adaptive radiation, coevolution and quantifiable selection pressures.

Within the present Volume, we have tried not only to provide some of the most modern concepts in the original descriptive science, on the sound base of which all further study must depend (papers by van der Pijl, by Stelleman, and by Vogel), but also to illustrate many new or potential developments towards an interdisciplinary, and often experimental approach. It could be said that a subject only matures when it contributes to other subjects, an overall, global wisdom being the final philosophical goal. Thus, the paper by Proctor indicates how pollination biology can be fitted into a more general ecological scheme.

In the case of the study by Corbet, a simple physiological technique gives a new insight into coevolutionary syndromes. The work by B. D. J. Meeuse is wholly physiological, showing clearly how the remarkable and distinct adaptations of sapromyophilous flowers provide a very suitable substrate for the investigation of sophisticated physiology. Equally suitable as a framework for further investiga-tions into insect behaviour and physiology are the studies by Kevan, by Brantjes and by Woodell. But the response of plants to aspects of insect behaviour can provide interesting and at times unexpected insights into taxonomic characters (often apparently "unadaptive") and incipient speciation, as shown in the paper by Eisikowitch.

Another major theme was the use of pollination biology in population biology and ecological genetics. Papers by Levin, by Beattie, and by Richards & Ibrahim, point out the role played by pollinators, and the importance of the careful study of this role, in population delimitation and subdivision. With the population now widely acknowledged as the basic evolutionary unit, the importance of such studies, apparently little recognised, can scarcely be overemphasised.

Populations do not only evolve randomly however; they do so in response to changing selection pressures, as Darwin (and now we have come full circle) first clearly demonstrated. It may be that the great difficulties genecologists have experienced in isolating clearly defined and quantifiable selection pressures acting on a genetic polymorphism, may be solved through studies of the differential pollination of varieties of flowers within a population. The papers by Kay and by Mogford show how this topic may develop, and provide useful reviews. It is also worth emphasis that all syndromes favouring cross-pollination, and thus the main objects of interest to pollination biologists, presumably arise in response to a requirement for heterozygosity, through both heterosis and the need for genetic variability in a heterogenous environment.

Pollination biology as a service to science is therefore a theme: but wide as its applications may be and however sophisticated it may recently have become, its roots still lie firmly in descriptive natural history, and here resides much of its charm. Much useful work can still be done with the minimum of money, equipment and expertise: indeed by the amateur. It is a subject with great amateur traditions (Darwin) and one in which important discoveries remain to be made by the patient and observant, asking the right questions. In population biology and ecological genetics, great advances can be made by amateurs, by insect-following for instance, or by studying flower phenology, or through studying visitors to flower colour polymorphisms, and showing resulting selection pressures (for instance in differential seed set).

Perhaps this is among the last subjects in which important, experimental, interdisciplinary advances remain to be made by the amateur. It is to the credit of a largely amateur body, the Botanical Society of the British Isles, that it should have arranged this conference, and obtained the support of the Linnean Society. Perhaps it has performed a limited service to amateurs in pointing out a little of the fascination, and importance, of the type of work that they could so readily and usefully undertake themselves.

Trends in research in pollination ecology

K. FAEGRI

University Botanical Museum and Garden, Bergen, Norway

This is a general review paper in which a wide range of topics of current interest are covered, and reference is made to developing techniques and concepts. Subjects include: distribution of anemophilous pollen in air; use of TV, UV, SEM, tape-recordings, electrophysiological and chemical techniques to investigate entomophily; fat oil attractants and perfume collectors; rendezvous attraction; new pollinator classes; energy budgets; theoretical concepts.

KEY WORDS:—pollination ecology—review—progress—techniques—concepts.

CONTENTS

ANEMOPHILY

Although this subject has little to do with the subject of the conference, it is perhaps of interest to start by mentioning the important data arising from observations made by Ogden, Raynor & Vormevik (1964). Snap-shot pictures of pollen distribution in air demonstrate great heterogeneities in both time and space. Using short-wave radar, Hardy & Ottersten (1968) have shown the importance of anemophilous pollen travel in rising turbulent air-masses (cells). Primary air-cells are a few hundred metres across, but composite cells are perhaps five times as large. This work has proved to be of great importance to the medical investigator of allergies as well as the pollination biologist.

Various studies in forests have rather surprisingly demonstrated that pollen transport by air currents inside the forest is negligible, with the influence of individual trees extending only over a 20–50 m radius. This raises a most interesting question about the fertility of isolated trees in self-incompatible or unisexual species, not least in tropical forest where such isolation may be the rule (and also incidentally about seed-dispersal by wind to give such isolated plants).

NEW OBSERVATIONAL METHODS IN BIOTIC POLLINATION
Visual

The introduction of colour TV techniques has permitted the observation and control of the UV reflecting component in flower colours, and thus the evaluation of insect colours, without being dependent on laborious, time-demanding and sometimes controversial photographic techniques using filters and films with unusual spectral sensitivities. However, for permanent recording video tapes are less satisfactory, being cumbersome to use and carrying poor resolution. TV techniques have pioneered the discovery of new classes of substances responsible for UV colours in addition to the carotenes and anthocyanins known from visual spectra.

Use of photomultiplication devices permit observation at natural light even at night and may become the ideal solution for observation of nocturnal pollination.

Also relevant here are studies of the pollen/stigma interaction using the SEM. These have helped to shed a good deal of light on the mechanism of self-incompatability; for instance Roggen's (1972) studies of films on stigmatic papillae and their breakdown by legitimate pollen.

Acoustic

Reference to Macior's (1968) work in distinguishing between flight and feeding (collecting) wing-beats of pollinators is relevant here. This used tape-recordings played back at various speeds and visual sound-trace analysis.

Electrophysiological

These include the use of antennograms to check for odour perception in insect pollinators. Such techniques enable the detection of antenna reaction on individual chemical substances, especially in connection with sexual attraction syndromes (Kullenberg, 1973, on *Ophrys*).

Chemical

Modern techniques, especially of microchromatography, have enabled progress to be made in the elucidation of the occurrence and nature of recognition substances involved in the pollen/stigma interaction. Some of these substances, presumably involved in human allergies, are easily leached out by water, resulting in allergogenic reactions in a few minutes. These substances must be located in outer part of the exine (presumably excreted from the gametophyte in gametophytic systems, but added externally from the tapetum in sporophytic systems). The role played by pollen kitt, apparently a complex mixture of protein and lipid, is not clear in this context, although it appears to be involved in the creation of callose plugs preventing pollen-tube penetration in at least some reactions. Thus perhaps it has a more active biochemical role than merely acting as a glue. At any rate some pollen loses the ability to germinate after a short wetting, presumably through the leaching of recognition substances (enzymes? antibodies?). Immersion in acetone has not resulted in the loss of germination ability, nor has deep freezing.

ATTRACTANTS

Traditionally, attractants to pollinators in flowers have been thought of as food (pollen, nectar), brood places, and later, sexual attraction. Not only sexual attraction, but also food attraction can be based on deceit. The list of attractants has been expanded by Vogel's (1974) discovery of two new classes: fat oil, and perfume. Fat oil is a typical, although previously unsuspected food attractant. As pollen contains variable amounts of fat, it is reasonable to assume that pollen-eating insect larvae possess enzymes capable of digesting fat. The problem lies in explaining the development of this syndrome. Thus, although the development of nectar can be explained as a development of the existing organisation of the plant (excreting excess carbohydrates in the transport system?), fat is a notoriously scarce substance in plants and its use as a lure in seed dispersal is a rare occurrence (olives, oil palm). The elaiosome attraction in myrmecophily is an interesting parallel.

Perfume-collectors fall within the sphere of sexual attractants. Some male hymenoptera produce perfumes themselves and distribute them into the environment ("male bumblebees are living perfume brushes"—Kullenberg, 1973). These perfumes are used to mark territory, for instance rendez-vous areas, well-known to occur in insects; to mark trails, "label their empties" etc. Perfume collectors do not produce perfume themselves, but collect plant perfumes to use in a similar manner, rather in the manner of aposematic insects.

Rendez-vous attraction is worth a further comment, for in these syndromes, the presence of one sex of an insect pollinator in the flower attracts the other sex, and many insects copulate in flowers (e.g. Umbelliferae). Such patterns are likely to improve the efficiency of pollination. However, it is not clear whether any cases are known of "pure" rendez-vous attraction, without the additional lure of food, or a brood place. Could *Protea* be an example?

Further light has also been shed on various types of deceit attraction. The two types known previously were sexual deceit, and food deceit based on the odour imitation of decaying proteins. However, in *Arum conophalloides*, the odour appears to mimic that of vertebrate skin, and Vogel has described how some mycetophilous gnats are attracted by a smell imitating that of fungi in a blossom.

NEW POLLINATOR CLASSES

For 150 years, the European-based study of pollination ecology has been myopically fixed on insect-pollination, although pollination by other groups has long been observed and correctly interpreted. Even among the insects, it was assumed that Hymenoptera, Diptera and Lepidoptera were of paramount importance, and occasional observations of pollination by other groups was written off as unimportant, as indeed most of them probably are. However, such pollinators as ants, especially in arid areas, mosquitos and pollen-eating butterflies and above all, beetles are now more fully evaluated.

Over the last 50 years, more attention has been paid to the role played by vertebrates in pollination. These have different problems to insects, having a protein requirement in their food, and through being homiothermic, requiring a steady energy intake, rather than being seasonal, as insects. As a result vertebrate pollinators are mostly restricted to tropical latitudes, although not necessarily to tropical climates (as in the afro-alpine zone). Alternatively, obligate vertebrate flower-feeders can migrate, although long distance migration is only known in humming-birds. Most vertebrate flower feeders have been correctly interpreted as nectar feeders only, but some of them, like the marsupial *Tarsipes*, brush-tongued lorikeets and the bat *Leptonycteris* show adaptations to total flower feeding, taking pollen protein as well as nectar from blossoms.

The vertebrate groups involved are small and able to reach flowers through climbing or flying; thus birds, marsupials, primates, bats, rodents and mammalian insectivores are principally concerned: all except the last have species known to be structurally adapted to flower feeding. Climbing pollinators have a lower energy budget than flyers, but tend to be less effective owing to smaller potential ranges.

Many problems remain in the identification and classification of pollinators. For instance, what pollinates geochoric flowers such as these of *Aspidistra* or some species of *Protea*? In some cases there are clear instances of sapromyophily (Vogel, 1959). In others, the pollinator is less certain, but there seems to be no reason why small vertebrates should not discover food and other amenities offered by flowers, use them, and adapt to them. The destruction of the flower is immaterial if they are pollinated and the gynoecium is not destroyed. Such more or less incidental blossom visits, which may or may not result in pollination, are observed in many animals apparently unadapted to anthophily, such as European birds utilising introduced ornithophilous blossoms, reptiles licking nectar out of flowers, etc.

ENERGY BUDGET

This is one of the most interesting of modern approaches. Although it has always been taken for granted that pollinators obtain food from flowers (excepting deceit attraction), this has only recently been discussed in relation to an energy budget (Heinrich, 1957). Theoretically, this approach is very important here, but the study has now been bedevilled by a lack of data, which require sophisticated experimental techniques.

There seems to be a primary distinction between functionally homiothermic and functionally heterothermic pollinators, the former group including some flying insects that are able to increase thoracic temperature above the ambient and are therefore not immobilised by lower temperatures. All flower visits require energy, and are energetically inefficient if expenses exceed gains. Negative items in the energy budget are the need to maintain a temperature differential between the body (for insects the thorax) and the air in cold weather; long flying distances between blossoms and between nest and foraging area; and the work entailed in the process of extracting nectar from the blossom. Crawling is (on a time basis) a more economic type of locomotion than flying, and bumblebees crawling on extensive inflorescences can let their thoracic temperature drop, thereby making the process so energy-effective as to be able to utilise small flowers with very little nectar.

Smallness of flowers in itself becomes an important ecological factor, as large pollinators with a large energy demand cannot afford to utilise these sources, which are thereby reserved for visitors with a low energy budget. Ants may be an important group to consider in this respect.

A distinction must be made between energy (nectar) collecting flights and pollen (or perfume) collecting, as pollen provides no energy to imagines. Thus, pollen collectors sip nectar occasionally ("to keep engine fuelled") or take nectar with them.

The consideration of energy budgets requires an examination of activity spectra. Flowers tend to be periodic (one or several openings) and pollinators show a corresponding periodicity. For these it is a question of "first come, first served", finding full nectaries and anthers. For example, an early morning visit may expend 0.8 calories in temperature loss, but be repaid by 1.5 calories. Although the calorie expense at a noon visit (higher temperatures) is very small, the reward is only 0.3 calories because of depletion of the nectar, thus the reward is as a whole so small that the visit is less efficient. Matinal bees can rob flowers of attractants, leaving very little for later pollinators (Baker, Cruden & Baker's, 1971, "minor pollinators"). However at the latter ambient temperature, no nectar is spent for maintaining temperature. All nectar collected represents a net gain.

The activity spectra of flowers and pollinators coincide, both on daily and yearly rhythms. Some of this correspondence may be due to meteorological parameters: temperature, light, humidity perceived by both. However, these cannot comprise all causal relations: periodicities are also partly endogenous, yearly or daily. On the other hand, flower formation and opening depend on a red/far red light signal which is probably without significance to insects, being outside of their spectral sensitivity. The concept of time memory seems doubtful.

The consideration of energy budgets throws much light on secondary phenomena. For instance, territorial hummingbirds spend much energy in territorial defence and patrol, and are dependent on clustered flowers giving great energy rewards for small expense. Non-territorial species (or individuals) may utilise scattered flowers ("trap-lining") providing a possibility for long-distance pollen dispersal. Long flights are more likely to occur with relation to energy independent pollen or perfume collecting than with nectar (energy) collection.

Also, the consideration of the origin of pollination and early adaptation of pollinators is aided by energy considerations.

PHYTOCOENOSIS RELATIONS

This is another new field of interest, concerning the total relationship of pollination activities within the syndrome. What do pollinators do when they are not actively pollinating the plants under consideration? Completely interdependent relationships, like those of *Ficus*, *Yucca* or *Pyrrhopappus/Hemihalictus* (Estes & Thorp, 1957) form closed systems independent of the coenosis in this respect. However, these are exceptions for the non-synchroneity of the life-times of pollinators and anthesis of their food plants makes it imperative that alternative food sources are available. Pollen of six different anemophiles was necessary to keep alive the bumblebee population of a salt-marsh (Pojar, 1974). The building up of a strong stock of pollinators by the utilisation of different sources of food early on is a well-known technique, above all in honey production, and similar relations seem to exist in nature.

Synergism is another coenosis effect: dominant flowers attract pollinators (Arnell, 1903); two subdominants with the same pollination syndrome reinforce each other's attractiveness (Macior, 1968) and may through synergism achieve a dominance effect which neither of them could have achieved alone.

Introduction of foreign pollinators may change competition conditions in a plant community, and thereby its composition—introduction of bees has been important in many parts of the world. The arrival of *Xylocopa* to the Galapogos islands furthered the establishment of non-native bee-pollinated plants (Linsley, Rich & Stephens, 1966). On the other hand, the selective removal within a biocoenosis of one link in a food-source chain may disrupt the whole system by depriving other flowers of a pollinator which at one period in its life-cycle was dependent on the absent food-source. Similarly, competition between plants is very important in agronomic fruit (seed) production, care being taken that pollinators are not diverted from the crop (*Taraxacum* in orchards; Alfalfa pollination).

It has been suggested that the attraction of the blossom of the agamospermous *Taraxaca* is also a means of competition by keeping pollinators away from competing plants and thereby reducing their seed-set.

Negative effects should not be overlooked. The effect of the red spot replacing the yellow nectar-guide (what are these colours in the bees' visual spectrum?) in old, dry *Aesculus hippocastanum* flowers is to keep insects away. Thereby, they are not diverted from the young, as yet unpollinated flowers, to older flowers that have been pollinated. The effect of keeping the spent flowers looking fresh is to maintain the general attractivity of the (morphologically very intricate) inflorescence.

THEORETICAL CONCEPTS

These have developed immensely under the stimulus of empirical discoveries, especially the concept of coevolution, which is still poorly understood. However, three major stages in the development of coevolution are currently recognised:

(1) Accidental utilisation without adaptation. Pollination effect uncertain.

(2) Adaptation without dependence. Pollination effect higher.

(3) (Inter)dependence. Pollination effect to 100%.

Stage (1) must have existed at a stage when sporophyll assemblages were not adapted to biotic spore transport, but exhibited some feature of interest to certain animals, probably the spores themselves. However, it is quite possible that pollination drops, vegetative nectar production through external sugar secretion analagous to modern extra-floral nectaries, may have occurred. The presence of other attractants is unlikely. The presence of food-bodies is possible, but the scarcity of these in modern systems renders this possibility unlikely.

Once the relationship between the pollinator (insect) and the plant has been established, further evolution is conceptually simple, through a "zig-zag" system in which each evolutionary step taken by the flower is followed by one by the pollinator and back again. This evolution is continuing at present: for example, the apparently recent tendency for sparrows to damage (and presumably pollinate) *Crocus* and other spring plants, and the pollination of garden *Puya* by European blackbirds. These examples illustrate quite effective recent associations in areas where the natural pollinator is absent. It will be interesting to see whether further coevolution will occur to increase the effectiveness for both the flower and the pollinator, and if so, at what rate. Other potentially new situations can be quoted which would repay careful watching, for instance the occurrence of humming-bird flower syndromes in plants cultivated or adventive away from the Americas. In view of the apparent universality of the attention of this syndrome to birds (Old World sunbirds, etc.), will new bird pollinators arise in these areas?

Dependence of the flower on a pollinator can be general or specific. General dependence is very common, and is due either to non-leaky self-incompatability or to rigorous heterogamy. Flowers showing general independence of pollinators can be classed either as autogamous or (at least facultatively) anemogamous. General dependence on flowers in animals is rarer than the converse, for most flower-visiting animals are able to sustain life through other media as well. The exceptions appear to be bees, some birds, higher butterflies and *Tarsipes*. In contrast most lower groups of insects and pollinating vertebrates develop from occasional to regular flower visiting in competitive situations.

Specific dependence of the flower usually leads to specific dependence in the pollinator to complete interdependence, as in *Ficus*, *Yucca*, *Tarsipes/Banksia*, and *Parkia*/bats. Such interdependence leads to ecological limitations of both partners based on the limitations of each, which can lead to the geographical limitation or extinction of one, due to catastrophe in the other.

In angiosperms, anemophily is probably secondary, although cases of revertance to entomophily have been described. The pollen of anemophiles is readily used by pollen collecting insects, but the overall effect on pollination is probably minor in many cases. The dry pollen grains of typical anemophiles, while still nutritious and therefore collected, adhere very badly to the body of insects and the transportation effect is therefore insignificant. However, the pollen of anemophiles (often very abundant) may serve to maintain a pollinator population which would not have been able to survive on the entomophilous flowers present. The problem of poisonous (to pollinators) pollen and nectar is hardly understood.

An increasingly popular field is the importance of pollination systems in taxonomy and evolution. Thus it has been shown on several occasions (Grant, 1971) that genetically viable hybrids may be unsuccessful due to poorer pollination. On the other hand, the establishment of an independent pollination syndrome may save a hybrid from being back-crossed and thereby establish it as an independent taxon. This may be one of the reasons for the explosive speciation in orchids. A study of the breeding system is of primary importance in population biology, but has rarely been carried out. For instance, clines are most commonly found in widely dispersed species with non-selective (wind) pollination systems. Distinctive localised ecotypes are more frequent in specifically entomophilous species with discrete local populations and restricted breeding systems. The extreme case of this is encountered in the "ethological" species in the orchids. It is worth nothing that the exceedingly important role played by pollinators and breeding systems in plant evolution still receives no mention in a recent paper entitled "Trends, priorities and needs in systematic and evolutionary biology", published in *Systematic Zoology* (*23*: 413–439).

REFERENCES

ARNELL, H. W., 1903. Om dominerande blomningsföreteelser i södra Sverige. *Arkiv för botanik, 1:* 287–376.

BAKER, H. G., CRUDEN, R. W. & BAKER, I., 1971. Minor parasitism in pollination biology and its community function: the case of *Ceiba acuminata. Bioscience, 21:* 1127–1129.

ESTES, J. A. & THORP, R. W., 1957. Pollination ecology of *Pyrrhopappus carolinianus* (Compositae). *American Journal of Botany, 62:* 148–159.

GRANT, V, 1971. *Plant Speciation.* New York: Columbia University Press.

HARDY, K. R. & OTTERSTEN, H., 1968. Two scales of convection in the clear atmosphere. *Proceedings of the International Conference on Cloud Physics, Toronto:* 534–538 (American Meteorological Society).

HEINRICH, B., 1957. Energetics of pollination. *Annual Review of Ecology and Systematics, 6:* 139–170.

KULLENBERG, B., 1973. Field experiments with chemical sexual attractants on aculeate hymenoptera males. II. *Zoon, Suppl., 1:* 31–42.

LINSLEY, E. G., RICH, C. M. & STEPHENS, S. G., 1966. Observations on the floral relationships of the Galapagos carpenter bee. *Pan-Pacific Entomologist, 42:* 1–18.

MACIOR, L. W., 1968. Pollination adaptation in *Pedicularis groenlandica. American Journal of Botany, 55:* 927–932.

OGDEN, E. C. RAYNOR, G. S. & VORMEVIK, J. M., 1964. Travels of airborne pollen. *Progress Report No. 5, New York State Museum Science Survey.*

POJAR, J., 1974. Reproductive dynamics of four plant communities of southwestern British Columbia. *Canadian Journal of Botany, 52:* 1819–1834.

ROGGEN, H. P. J. R., 1972. Scanning electron microscopical observation on compatible pollen-stigma interaction in *Brassica. Euphytica, 21:* 1–10.

VOGEL, S., 1959. Organographie der Blüden Kapländischer Orchideen. I-II. Akademie der Wissenschaften und der ditteratur (Mainz). *Abhandlungen der Mathematischnaturwissenschaftlichen Klasse, 1959:* 265–532.

VOGEL, S., 1966. Parfümsammelnde Bienen als Bestäuber von Orchideen und *Gloxinia. Österreichibotanische Zeitschrif, 113:* 302–361.

VOGEL, S., 1973. Fungus mimesis of fungusgnat flowers. In N. B. M. Brantjes & H. F. Linskens (Eds), *Pollination and Dispersal:* 13–18. Publ. Dept. Botany, Nijmegen:

VOGEL, S., 1974. Ölblumen und ölsammelnde Bienen. *Trop. subtrop. Pflanzenw., 7:* 283–547.

A more complete bibliography will appear in the forthcoming 3rd edition of Faegri & van der Pijl, *Principles of Pollination Ecology.*

Sensory responses to flowers in night-flying moths

N. B. M. BRANTJES

Botanisches Laboratorium, University of Nijmegen, the Netherlands

Very small moths (e.g. Geometridae) prefer to drink from the same species of flower, as do mosquitoes. Both can orientate towards the source of flower odour.

In large moths (Sphingidae and Noctuidae) odour acts as a stimulus for the release of the feeding behaviour, both in trained and in untrained moths. This behaviour is recognisable by a characteristic flight pattern, "seeking flight", and by a positive reaction to white or coloured objects.

Sphingidae (*Manduca sexta, Deilephila elpenor*) react specifically to odours. Noctuidae (*Cucullia umbratica, Hadena bicruris, Autographa gamma*) react to a wide variety of flower odours.

Both moth families are able to orientate towards flowers by odour only, but mostly they use visual cues in combination with odour. Positive anemotaxis, released by scent, seems to occur in Sphingidae. Wind tunnel experiments proved the existence of this means of orientation in Noctuidae. In standing air, orientation over a short distance, towards a source of scent is possible for *D. elpenor* and for several Noctuidae species. The attractiveness of the source of scent to *H. bicruris* depends on concentration. In the presence of an effective flower odour, Sphingidae and Noctuidae can direct their flight towards scentless white and coloured objects. The Noctuidae do not land on these. However, Sphingidae direct the proboscis towards the flower entrance with the aid of visual cues. In contrast to this, Noctuidae use differences in odour quality on the flower surface to locate the flower opening.

The ovipositing behaviour of *H. bicruris* is connected with the drinking behaviour, which results in pollination and development of the ovules, the food for the larvae.

The moth selects the pistillate flowers of *Silene alba* for oviposition. This selection is based on differences in petal odour between pistillate and staminate flowers. The probability for oviposition selection is, compared to flowers without eggs, lower in flowers that have already received an egg.

Several *Hadena* species have a specific preference for oviposition in the flowers of Caryophyllaceae species. The evolution of this relationship is discussed.

KEY WORDS:—Sphingidae—Noctuidae—odour—anemotaxis—*Hadena*—*Silene*—ovipositing.

CONTENTS

INTRODUCTION

The following discussion about the sensory responses of night-flying moths focuses on one of the chemical senses: the perception of odours, or smell. The impact of odour has been analysed by observing the behaviour of moths in the field and using laboratory experiments. In this paper, I am concentrating on moth behaviour, and will ignore the consequences to the plants of this behaviour: the pollination. It is well known that adult Lepidoptera visit flowers mainly for feeding by drinking the nectar. But there are also a few species that use the flowers in addition as substrate for oviposition. This rare combination of two behaviour types will be included at the end of this paper. Because not all kinds of moths show the same type of behaviour, the role that odour plays might vary. Therefore, I am differentiating moths into three broad categories:

Very small moths
Noctuidae
Sphingidae

VERY SMALL MOTHS AND DIPTERA

Many families of Lepidoptera have small representatives that visit flowers. Because they are nocturnal, these tiny moths are often overlooked and, therefore, observations of their behaviour are scarce. The small moths mostly approach flowers by flying upwind. This suggests that they may be flying along the odour trail coming from these flowers. If netting is put as a screen over plants with flowers, the small moths still land and pierce with their proboscis through the holes on the leeward side (Brantjes & Leemans, 1976). This indicates an odour triggered response resulting in the uncoiling of the proboscis, and the directing of it towards the source of the odour (for sight and feeling can be eliminated).

Mosquitoes have a striking parallel in behaviour with small moths. Both visit the same flower species for drinking, and often can pollinate them. The mosquitoes of both sexes land on the leeward side of the netting screen over flowers, just as the moths do (Brantjes & Leemans, 1976). This indication of orientation with the aid of flower odour was finally proved by luring mosquitoes with flowers hidden in boxes. Other small Diptera, e.g. Empididae, also visit and pollinate the same flowers as small moths.

BEHAVIOUR OF LARGE MOTHS

In order to study how the other two types of moths (Sphingidae and Noctuidae) react to flower odours, we have to distinguish five elements in their flower visiting behaviour:

1, activation;
2, orientation;
3, landing;
4, localisation of the nectar;
5, oviposition (only for a few species).

Activation

Flower odour is the activator for the start of feeding behaviour. In an experimental cage with moths in flight, both Sphingidae and Noctuidae almost always show a prompt change of behaviour after introduction of flower odour into the air (Brantjes, 1973, 1976b). This change in behaviour has two characteristics: the flight pattern; and the reaction to colours.

In an odourless atmosphere, the moths fly slowly around, mostly in the top of the cage. After introduction of odour, this pattern of flight changes immediately: the moth shows an erratic flight with many sudden drops and slow rises. This is named the "seeking flight". It fits the description by Knoll (1922) and Schremmer (1941) of the "Nahrungflug" of (respectively) *Macroglossum stellatarum* (Sphingidae) and of *Autographa gamma* (Noctuidae). Knoll did not discover the key-factor for the seeking flight, but Schremmer observed that flower odour provoked this.

In the absence of odour the moths avoid flying against coloured or white objects. Therefore, they can see these, but the objects do not seem to have a special meaning to the moths. However, as soon as flower odour enters the cage, the moths start to approach all kinds of coloured or white objects. Under the influence of odour the reactions to these visual stimuli change. Activation of the feeding behaviour is independent from learning. Odour is necessary as a trigger both to moths that visited flowers the previous night, and to moths that have had no training with flowers. Sphingidae seem to react specifically to odours, whereas Noctuidae react to a wide spectrum of flower odours. This odour-activation of feeding behaviour might also function as a primitive means of locating flowers. A moth cruising over long distances will be checked, and will perform a seeking flight after entering a cloud of perfume. The odour will be an "arrestant".

Orientation in still air

I extracted flower-odours with a quick evaporating solvent (heptane) and applied the extract on strips of absorbent paper. After evaporation of the solvent, these strips were introduced into an experimental cage where a moth was active. At this point, the moth landed on the strip and scanned the surface with its antennae and proboscis. After simultaneous introduction of several strips, each with a different concentration, the number of landings on each strip was found to be proportional to the odour concentration (Brantjes, 1976b). The moths selected the strips with the highest concentration of odour and thus, apparently, use concentration differences for orientation.

To achieve this result, the moths have to perceive differences in odour concentration spaced wider than the distance between their antennae tips. With an increase of the distance from the odour source, this difference decreases. Therefore, concentration directed orientation is only possible over a limited distance. In a glasshouse this distance was less than 3 to 5 m for several Noctuidae (*Hadena bicruris, Cucullia umbratica, Autographa gamma, Plusia chrysitis*).

Sphingidae can also orient toward the highest odour concentration. I tested this for *Deilephila elpenor* in an experimental cage with flowers of *Lonicera periclymenum* hidden in a box, with one opening on the underside (Brantjes, 1973). The hawkmoth hovered in a pendulous flight at the opening, landed on the brim and, as a flashphoto shows, stretched its proboscis into the opening.

Orientation in wind

There are two ways a moth might orient itself along a scent trail formed by wind. The first, by following the edge of the trail, is very unlikely to occur because of the turbulence in the air. The second is through optomotor anemotaxis induced by odour, as Kennedy & Marsh (1974) demonstrated for a moth reacting to a pheromone. A similar positive anemotaxis can also be induced by floral odours: I tested this in a wind tunnel, where *Hadena bicruris* moths flew randomly around. Within ten seconds after the introduction of flowers into the airflow, upwind from the compartment with the moths and out of sight to them, all moths flew upwind and landed on the screen of metal gauze where the air came through. The moths pierced with their proboscis through the holes of the gauze. Less than a minute after removal of the flowers, the moths started flight again. The anemotactic reaction could be elicited several times with the same moths. Because the odour was mixed thoroughly in the wind tunnel, the moths could not orient towards a higher concentration or along the edge of the odour trail. The flower odour induced an anemotatic flight, in this representative of the Noctuidae.

Tinbergen (1958) observed the upwind approach of *Sphinx pinastri* toward hidden flowers of *Lonicera periclymenum*. This indicates that Sphingidae can also orient anemotactically along a scent trail.

Anemotactic orientation is effective over long distances as long as the odour is perceived by moths, but concentration directed orientation can become utilised after the moth comes very close to the flowers by flying upwind.

Visual versus olfactory senses

Orientation by smell is sufficient. This can be seen from experiments in absolute darkness where Noctuidae visited flowers. The importance of orientation by smell for moths is also indicated by the existence of moth flowers without visual attraction, where petals and sepals reduced to green osmophores (e.g. *Narcissus viridiflorus*, described by Vogel & Müller-Doblies (1975) and *Silene otites* (Brantjes & Leemans, 1976)).

However, we must be careful not to exaggerate the value of flower odour as a means of orientation. Experiments reported in this paper up to this point concern the maximal, potential importance of odour to the moth. But, in the field visual cues are always present and are used by different moth species to a different extent. For instance, anemotaxis is based on visual perception of the apparent movement of the ground pattern. Secondly, coloured and white objects are perceived and often approached by moths after their feeding behaviour has been activated. Knoll (1925) observed that Sphingidae did not discriminate between scented and scentless objects. In contrast to this, I have observed that *Deilephila elpenor* works for a longer time with the proboscis (in an attempt to reach the nectar) on scented objects than on scentless objects presented simultaneously. Here the presence of odour might enlarge the motivation, rather than have an orientational function. In *Autographa gamma*, a shift of orientation towards the visual occurs after successive visits. Schremmer interpreted this as the consequence of a learned association between visual stimuli and the reward (nectar).

Landing

In *Hadena bicruris*, even individuals that never visited flowers before directed their flights toward scentless objects, but usually without landing.

A stimulus is necessary for landing, and for the unrolling of the proboscis. Moths land without hesitation on locations with flower odour. Therefore, this odour seems to be the stimulus for landing. The threshold value might be lowered by several factors. One can be training, as Schremmer observed with *A. gamma*. Another factor might be starvation—sometimes *Hadena bicruris* that had not been fed for days landed on scentless objects and probed these with their proboscis. Flower odour can also motivate Sphingidae to land and to unroll the proboscis, as shown by a photo of *Deilephila elpenor*, after its approach to a hidden source of scent.

Another stimulus for the unrolling of the proboscis in several species is the stimulation of the tarsal senses with sugar solution or with flower petals (Schremmer, 1941; Brantjes, 1976a). Flashphotos, taken at the moment of landing on a scentless resting place, showed that a variety of moths partly uncoiled their proboscis just before or at the moment of touching the landing place, and these spiralised again immediately afterwards (Callahan, 1965). The onset of despiralising might, therefore, form a part of the landing action.

Localisation of the nectar

After approach and landing, the proboscis tip has to enter the opening of the flower. The difference in behaviour between Sphingidae (hovering) and Noctuidae (landing) has consequences for the discovery of the flower entrance. Hovering causes air turbulence and this makes the use of odour differences very unlikely. Knoll showed that *Deilephila livornica* (Sphingidae) directs its proboscis visually.

In contrast to the Sphingidae, the Noctuidae direct the proboscis with the aid of differences in odour quality over the flower surface. A first indication can be the vivid movement of the antennae over the flower surface after landing. Proof can be found in the behaviour of all species I tested with flowers in glass tubes. The moths directed their proboscis on the netting over the opening of the tubes and not to the flowers behind the glass. Also, *Hadena bicruris* in the wind tunnel directed their proboscis towards the inflowing odour.

Perceptible odour differences to moths, qualitative or quantitative, apparently exist on the surface of flowers of *Silene alba*. The ligula is functional in directing the proboscis towards the flower opening, as shown by the behaviour of *Cucullia umbratica* (Brantjes, 1976a) and *Hadena bicruris*. After the ligula had been cut from the flowers, *Hadena bicururis* took an average of two seconds, instead of one, to insert its proboscis into the flower tube. After removal of the ligula, *Hadena bicruris* scanned the whole surface of the petal with antennae and proboscis tip. But when the ligula is present, the moth restricts the probing to this part, even on isolated petals. There, in contrast to Sphingidae, Noctuidae may utilise odoriferous nectar guides. Thus, the observation (Vogel, 1962) of localised odour production in *Platanthera bifolia* can be an argument for regarding this flower as a Phalaenophile rather than a Sphingophile.

OVIPOSITION

This ends the role played by the odour in flower visits, and this would also be the end of this paper if it were not for one genus, *Hadena* which continues its visit to a flower by laying an egg. *Hadena bicuriris* does this exclusively inside pistillate flowers of *Silene alba*. The connection of the nectar-drinking and of the oviposition has two aspects. Firstly, both activities can be performed in the same flower. Secondly oviposition never occurs without prior drinking. The advantage to the moth is clear: the flower becomes pollinated during the drinking and so the larvae get their food, developing ovules, prepared at the right time.

The factors through which *Hadena bicruris* selects flowers to lay eggs in were studied experimentally (Brantjes, 1976c). It was of especial interest to discover how the moth discriminates between pistillate and staminate flowers, which even many biologists can't tell apart! To determine which parts of the flowers were involved in the discrimination, I designed an artificial flower, the pipette flower, in which isolated flower parts in different combinations could be inserted. Only petals were effective in inducing the laying of an egg. The stimulus to the moth is the petal odour, which I hope to identify this summer. The difference between the odour of pistillate and of staminate petals is quantitative. Staminate flowers can receive eggs if the flower odour of pistillate flowers is reduced by removal of several petals. The ligula also contributes to the odour that induces the moth to lay its eggs.

Schremmer demonstrated that *Autographa gamma* can form associations between flower odour and nectar. *Hadena bicruris* also does so. However, this association which is formed during drinking, hardly influences the selection of the substrate for oviposition. Therefore, oviposition is linked to the drinking behaviour, but is governed in a different way. The preferences are innate, just as we have seen for the activation of the drinking behaviour by flower scent.

Variation is a characteristic of nature that has often been denied by social philosophers. Variation in odour would cause some flowers to be more likely to be selected for oviposition. In the botanical garden, there are fewer such flowers, bearing two or more eggs, than would be expected, even with random flower selection. The moths spread their eggs very evenly over the available flowers (Brantjes, 1976d). Last year I also observed this equalisation in the field. From this I conclude that moths are able to observe whether a flower contains an egg or not. However, ovipositions from the previous night are not perceived. Therefore, it is unlikely that the eggs themselves, or physiological changes in the flower are relevant. How the moth observes the previous oviposition in the flower is not yet clear. A scent mark given off during oviposition is a good possibility.

EVOLUTION OF THE RELATIONSHIP

The well adapted behaviour of *Hadena bicruris* in its relation with *Silene alba* shows parallels with *Tegiticula* in its relationship with *Yucca*. Yet the relationship is not one to one. Several other Caryophyllaceae are acceptable as larval food and these are also accepted as secondary substrates for oviposition in the absence of *Silene alba*, in a distinct order of preference. Many more *Hadena* species eat seeds of the Caryophyllaceae, and show different arrays of oviposition preferences, which can be considered to be adaptive.

It is tempting to talk about coevolution. But for coevolution it is not sufficient that the moth is adapted towards its foodplant. It is also necessary to show that the plant benefits from specialised pollinator visits, which in a diclinous ento-mophilous species such as this may well be the case. However, the specialisation of several *Hadena* species on the different Caryophyllaceae could also be the result of adaptive radiation of seed-eaters. Before we can decide on coevolution, the respective influences of the pollinators and seed-destructors on the evolution of the Caryophyllaceae has to be established.

One interesting example appeared last year in Africa (Stirton, 1976). There, several representatives of the subfamily Hadeninae eat seeds from flowers they also pollinate. But in that case the moths specialised on the Liliaceae and Amaryllidaceae for both activities (Stirton, pers. comm.).

REFERENCES

BRANTJES, N. B. M., 1973. Sphingophilous flowers, function of their scent. In N. B. M. Brantjes & H. F. Linskens (Eds), *Pollination and Dispersal:* 24–46. Nijmegen: Publ. Dept. Botany.

BRANTJES, N. B. M., 1976a. Senses involved in the visiting of flowers by *Cucullia umbratica* (Noctuidae, Lepidoptera). *Entomologia experimentalis et applicata, 20:* 1–7.

BRANTJES, N. B. M., 1976b. Riddles around the pollination of *Melandrium album* (Mill.) Garcke (Caryophyllaceae) during the oviposition by *Hadena bicruris* Hufn. (Noctuidae, Lepidoptera), I. *Proceedings. K. Nederlandse akademie van wetenschappen, Ser. C, 79:* 1–12.

BRANTJES, N. B. M., 1976c. Riddles around the pollination of *Melandrium album* (Mill.) Garcke (Caryophyllaceae) during the oviposition by *Hadena bicruris* Hufn. (Noctuidae, Lepidoptera), II. *Proceedings K. Nederlandse akademie van wetenschappen, Ser. C, 79:* 125–141.

BRANTJES, N. B. M., 1976d. Prevention of superparasitation of *Melandrium* flowers (Caryophyllaceae) by *Hadena* (Lepidoptera). *Oecologia, 24:* 1–6.

BRANTJES, N. B. M. & LEEMANS, J. A. A. M., 1976. *Silene otites* (Caryophyllaceae) pollinated by nocturnal Lepidoptera and mosquitoes. *Acta botanica neerlandica, 25:* 281–295.

CALLAHAN, P. S., 1965. A photoelectric-photographic analysis of flight behaviour in the corn earworm, *Heliothis zea*, and other moths. *Annals of the Entomological Society of America, 58:* 159–169.

KENNEDY, J. S. & MARSH, D., 1974. Pheromone-regulated anemotaxis in flying moths. *Science, 184:* 999–1001.

KNOLL, F., 1922. Lichtsinn und Blütenbesuch des Falters von *Macroglossum stellatarum. Abhandlungen der Zoologisch-botanischen Gesellschaft in Wien, 12:* 127–378.

KNOLL, F., 1925. Lichtsinn und Blütenbesuch des Falters von *Deilephila livornica. Zeitschrift für vergleichende Physiologie, 2:* 329–380.

SCHREMMER, F., 1941. Sinnesphysiologie und Blumenbesuch des Falters von *Plusia gamma* L. *Zoologische Jahrbücher, Systematik, 74:* 375–435.

STIRTON, C. H., 1976. *Thuranthos:* notes on generic status, morphology phenology, and pollination biology. *Bothalia, 12:* 161–165.

TINBERGEN, N., 1958. *Curious Naturalist.* London: Country Life Ltd.

VOGEL, S., 1962. Duftdrüsem im Dienste der Bestäubung. *Abhandlungen, Mathematisch-naturwissenschaftiche Klasse. Akademie der Wissenschaften und der Literatur, Mainz, 10:* 599–763.

VOGEL, S. & MÜLLER-DOBLIES, D., 1975. Eine nachtblütige Herbst-Narzisse, Zwiebelbau und Blütenökologie von *Narcissus viridiflorus* Schousboe. *Botanische Jahrbücher für Systematik, Pflanzengeschichte und Pflanzengeographie, 96:* 427–447.

Bees and the nectar of *Echium vulgare*

SARAH A. CORBET

Department of Applied Biology, University of Cambridge

The nectar of *Echium vulgare* shows marked fluctuations in volume and in sugar concentration from hour to hour and from day to day. These changes are interpreted in terms of losses or gains of sugar or water due to secretion by the plant, removal by bees, and evaporation or condensation. Diel patterns of visits by social bees can be related to these changes in the quality and quantity of nectar in the flowers. Temporal fluctuations of this order are probably widespread in the nectars of flowers, and their ecological and evolutionary implications are considered. One practical consequence is the difficulty of expressing these rapidly-changing properties of nectar in a way that permits valid comparisons between and within species.

KEY WORDS:—bees—nectar—*Echium*—diel—sugar—*Sinapis*.

CONTENTS

"The honey extracted from the flowers is the nectar which they enclose, and which was so much boasted of by the ancients, who formed from it the celestial beverage of their gods, to which they gave the name of ambrosia." (Huish, 1817).

INTRODUCTION

Nectar, because it is a fuel whose energy content can be measured quickly and easily in the field, is a superbly quantifiable resource that has formed the basis for some elegant studies of such ecological principles as foraging strategies (e.g. Heinrich, 1976a) and competition (e.g. Wolf & Hainsworth, 1975; Feinsinger, 1976; Heinrich, 1976b). To pursue the ecological adventures of energy derived from nectar sugar, it is necessary to know something about the fluctuations in the availability of that resource in flowers in the field. One needs to know, for instance, how many measurements of concentration are necessary to characterise the nectar in a given species. And does it matter at what time of day the samples are taken? How much does weather influence concentration? How can one assess a flower's daily production of energy available to foraging insects? Can this information be derived from studies of nectar production in flowers that are protected from insect visits?

I will illustrate some of the changes that can be expected in floral nectars by tracing a day in the life of two species of flowers, focusing on fluctuations in the properties of their nectar. For the purpose of this presentation I shall take a frankly ambrosiocentric view: nectar is not the only thing that matters to insect visitors, but it can matter a lot in some flowers, and I think it deserves more attention than it is sometimes given. I hope this exercise will show that there are very important changes in the nectar of flowers in the field; that these changes may be easily measurable and (perhaps less easily) predictable; and that insect/flower relationships may often be more fully interpreted if changes in the amount and composition of nectar are taken into account.

CHANGES IN THE VOLUME AND CONCENTRATION OF NECTAR

This work is part of a study of the community of insects and other animals associated with the flowers of viper's bugloss (*Echium vulgare* L.) in the Breckland area of Norfolk. A crop of white mustard, *Sinapis alba* L., was also examined. I collected nectar in microcapillary tubes and measured the volume per flower; and then at once measured concentrations of sugar with a refractometer modified by the makers (Bellingham and Stanley, Tunbridge Wells) for small volumes (0.5 μl) of liquid. Concentrations are expressed as the equivalent percentage of sucrose by weight. This enabled the absolute quantity of sugar per flower to be determined. Shade temperatures were measured with a fine thermocouple, 50 cm above the ground close to the *Echium* patch.

Table 1. Evidence that the total nectar yield of protected flowers is affected by withdrawal of nectar (from Raw, 1953 and Boëtius, 1948)

Species	Duration of experiment (h)	Mean weight of sugar yielded by flowers sampled:		once (mg)
Rubus idaeus	36	3 times	8.2	5.7
Rubus fruticosus	24	twice	7.2	3.9
Echium vulgare	24	twice	3.0	0.6

One point of technique needs special comment. In this study I used flowers that were exposed to insect visits, whereas most field studies of nectar production are based on flowers from which insects have been excluded, often by gauze or paper bags or by a cage. The use of unprotected flowers has advantages as well as disadvantages. An obvious disadvantage is the impossibility of assessing the total quantity of sugar secreted into a reservoir from which insects remove unknown quantities at unknown intervals. But that problem is not solved by protecting the flowers. There is reason to believe that protected flowers do not give valid estimates of total nectar production either, possibly because they constitute a reservoir from which nectar is removed in unknown quantities at unknown intervals by reabsorption into the nectaries. Table 1 summarises some evidence that in *Rubus* (Raw, 1952) and *Echium vulgare* (Boëtius, 1948) flowers from which nectar is periodically removed yield, in total, more nectar (containing more sugar) than do flowers which have not been disturbed over the same period. Reabsorption is probably at least partly responsible for this discrepancy (Boëtius, 1948), and it has been demonstrated for other species by Lüttge (1962). Protected flowers often differ from those exposed to insect visits in other ways too: the microclimate inside them may be much modified (Geiger, 1965).

I used unprotected flowers because my programme was designed to explore the pattern of changes in the nectar resources available to insects in the field.

Each flower was sampled once only and then marked so that it could be avoided on future sampling occasions. Flower age was not taken into account, because although rate of secretion varies with flower age in *E. vulgare* (Boëtius, 1948), flower-opening is evenly spread through the day in this species (and in *E. plantagineum*; Nuñez, 1977), so that proportions of flowers in the various stages of anthesis should vary little through the day.

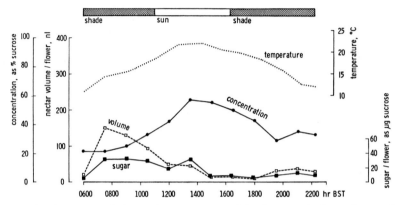

Figure 1. The time-course of changes in temperature and in the concentration, volume and sugar content of the nectar in flowers of *Echium vulgare* on 5 July 1975. Each record is based on a sample of at least 20 flowers.

The changes in the concentration and volume of the nectar of *Echium* proved to be well-marked: the amount and composition of the nectar varied from hour to hour and from day to day. Figure 1 shows changes in *Echium* nectar on a fine day, in weather chosen to present the simplest case. It was a calm sunny day and the relative humidity at the nearest meteorological station (5.6 km away) was 3⁹% at 10.00 hrs BST. As Boëtius (1948) commented, *Echium* shows such changes unusually clearly. But it is not unique; changes like this are probably widespread and have been demonstrated in, for instance, orange flowers (Vansell, Watkins & Bishop, 1942); *Tilia* (Park, 1929; Kleber, 1935); and *Sinapis* (Boëtius, 1948). My own observations on *Sinapis alba*, made for comparative purposes on the same days as the *Echium* records, show such changes too.

THE CAUSES OF THE CHANGES

These changes reflect a situation that is complicated by the interaction of a number of factors that influence the amount and concentration of nectar present in the flowers at any one time: the activity of the nectaries (secretion or reabsorption); equilibration with the humidity of the air (evaporation or condensation); and removal of nectar by insects (Table 2). It is convenient to begin by considering the total amount of sugar. The only thing that can increase the absolute quantity of sugar per flower is secretion, so at times when the amount of sugar per flower increases secretion *must* be happening (and it *may* be happening at other times too). So Fig. 1 shows secretion at 06.00 to 07.30 hrs and again at 18.00 to 21.00 hrs. (The increase in sugar quantity at 12.00 to 13.30 hrs is not substantiated by other relationships.)

Table 2. The effects of various factors on the concentration of nectar and on the amounts of sugar and water per flower

Effect of	on sugar concentration;	and on amount per flower of	
		sugar	water
Equilibration with air humidity:			
condensation	dilution	no effect	gain
evaporation	concentration	no effect	loss
Activity of nectary			
secretion	approaches 15–35%	gain	gain
reabsorption	? no effect	loss	loss
Insect visits	no effect	loss	loss

Next, we may consider concentration. Insect visits affect volume but they are unlikely to have a significant effect on the concentration of the nectar they leave behind (except indirectly, in so far as the small amount of nectar that remains will equilibrate faster with the humidity of the air than a larger volume would do). Any changes in concentration must therefore be due to the interaction of two effects: equilibration with the humidity of the air; and secretion or reabsorption by the nectaries. The relative importance of these two effects can be assessed by seeing how closely concentration follows changes in microclimate. Probably the key variable here is vapour pressure deficit (Williams & Brochu, 1969), but in the absence of records of vapour pressure deficit I have used records of temperature which, over this range, is nearly linearly related to vapour pressure deficit for a given absolute humidity. The curve of sugar concentration shows a close relationship with the curve of shade temperature with a delay of, in the case of *Echium*, about half an hour. In Fig. 2 the concentration is plotted with the temperature half an hour earlier, and it is clear that whereas concentration follows temperature closely for much of the day, there are two periods, one in the morning and one in the evening, when the concentration is closer to a value of about 20–35% than the temperature curve would lead one to expect. Those times coincide with the periods when a rising sugar quantity implies that secretion is taking place. The conclusion that *Echium* secretes its nectar at a concentration of about 20–35% sugar fits reasonably well with Huber's (1956) conclusion that in a range of plant species newly-secreted nectar contained about 12–34% of sugar.

In summary, then, this day's records of *Echium* show that nectar was secreted in the morning and evening at a sugar concentration of about 20—35%; and that after secretion the concentration of the nectar rose during the day, following changes in shade temperature with a delay of about half an hour. Post-secretory changes of this kind do seem to affect insects: Vansell *et al.* (1942) found that orange nectar was secreted with a sugar concentration of about 14–16%, and that only when the weather was dry enough to concentrate the nectar to about 30% sugar did honeybees visit the flowers. Kleber (1935) found that honeybees stopped visiting lime flowers when the nectar dried up in the afternoon but resumed their visits when the nectar became more dilute in the evening.

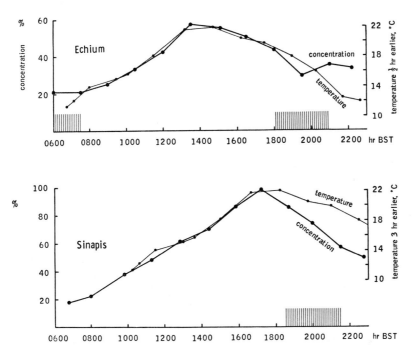

Figure 2. Changes in the concentration of sugar in the nectar of *Echium* (above) and *Sinapis* (below) in relation to the shade temperature half an hour earlier (for *Echium*) or three hours earlier (for *Sinapis*) on 5 July 1975. Hatched blocks indicate periods when an increase in the quantity of sugar per flower implies that secretion is occurring (Figs 1 and 3).

The records of changes in the nectar of white mustard, *Sinapis alba*, made on the same day as the *Echium* records, can be analysed in a similar way, and show some parallels and some differences (Fig. 2). Here again, concentration followed temperature, but a given change in temperature was associated with a greater change in nectar concentration in *Sinapis* than in *Echium*; the peak concentration was much higher in *Sinapis* than in *Echium*, so that the nectar became very viscous; and the concentration was related to the temperature three hours earlier, not half an hour earlier as in *Echium*. In *Sinapis*, as in *Echium*, a period of secretion in the evening after 18.00 hrs was indicated both by the timing of increase in sugar quantity (Fig. 3) and by the departure of the concentration curve from the temperature curve; but if there was a period of secretion in the morning, records did not begin early enough to show it. The graph does not show at what concentration *Sinapis* nectar was secreted, although this must have been at a concentration below about 50%.

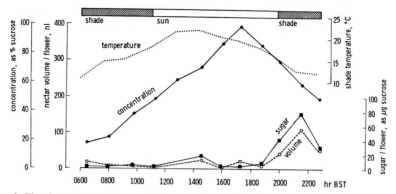

Figure 3. The time-course of changes in temperature and in the concentration, volume and sugar content of the nectar in flowers of *Sinapis alba* on 5 July 1975. Each record is based on a sample of at least 5 flowers.

The apparent importance of air humidity in these cases would lead one to expect that on a humid day the pattern of change in nectar concentration would be quite different; and, because of the effects of weather and concentration on insect visits, the pattern of change in volume and sugar content of the nectar would be quite different too. That is indeed the case for *Echium* and *Sinapis*. Three days after the experiment just described, the weather was overcast all day, and the relative humidity was high (70% at 15.00 hrs BST at the nearest meteorological station). On that hot and humid day, the concentration of *Echium* nectar remained nearly constant at about 30–35% all day, bee visits were fewer, and the nectar volume and sugar content were higher. No doubt environmental effects on rates of secretion or reabsorption are responsible for some of these differences, but weather also affects post-secretory changes in the nectar as well as insect visits.

SOME CONSEQUENCES OF CHANGES IN THE CONCENTRATION OF NECTAR

The concentration of sugar in nectar influences visitors to flowers, and there is some evidence that different groups of flower visitors differ in their preferred concentration ranges. Percival (1965) lists concentration measurements for a range of flower species that she has classified by visitor type, and by rearranging her list one can derive an approximate sequence of groups of flower-visitors, in increasing order of preferred concentration range (Table 3). The lower value in each range

Table 3. Sugar concentrations in the nectar of flowers of various pollination types (data from Percival, 1965)

Type of pollinator	Concentration range, %	Number of species
moth	8–18	2
bat	14–16	2
bird	13–40	7
butterfly	21–48	2
honeybee + bumblebee	10–74	24
(short-tongued bees?)	(higher?)	(—)
(short-tongued flies?)	(higher still?)	(—)

may represent dilute nectar samples, say, in the morning when it was plentiful enough to be easily measured but too dilute for visitors to take it. Concentrated nectars may be under-represented because their low volume and high viscosity make sampling very difficult, and perhaps that is why Percival includes no fly flowers (except primroses, whose visitors, *Bombylius*, have long tongues and may take their nectar weaker than most flies do). Probably if more concentrated nectars could be included in the table the sequence would end with short-tongued flies. This sequence makes sense if viscous solutions cannot be quickly sucked by insects with long tongues; and Betts (1930) has shown that honeybees' sucking rates decline markedly when the concentration of the sugar syrup on which they are fed exceeds about 50–60%. Simpson (1964) showed that honeybees taking up such concentrated solutions diluted them a few per cent by spitting into them a watery saliva from the labial glands (Simpson & Reidel, 1964). Lepidoptera, too, can make concentrated sugar solutions more easily potable by spitting labial gland secretion into their drink (Stober, 1927; Robertson, 1974). Both Lepidoptera and bees seem to feed on only moderately concentrated nectars in the field.

Short-tongued flies, on the other hand, take nectar at high concentrations (Elton, 1966) and readily feed on crystalline sugar. They do this by spitting on their food and lapping up the solution (Hansen-Bay, 1976). It may be their willingness to spit and lap that enables flies to exploit the dry nectar of open flowers abandoned by bees in the heat of the day.

In considering the evolutionary significance of nectar concentration, I would like to shift emphasis. It is conventional to regard the temporal patterns of change in nectar concentration as incidental correlates of floral types that evolved in relation to other selective forces. Let us instead regard flowers as sophisticated gadgets for dispensing nectar at the right concentration at the right time. If there is a degree of concentration specificity among pollinators, there will be a selective advantage for flowers whose morphology and secretory periodicities interact with the local climate in such a way as to increase the likelihood that the nectar will be at an appropriate concentration at a time of day when a suitable pollinator is active. I shall consider this idea in relation to two extreme flower types: open flowers and deep tubular flowers.

Open flowers are typically visited by what Allen (1891) calls "the insect riff-raff": short-tongued flies and beetles. Probably the preferred concentration range for these visitors is generally considerably higher than the concentration at which nectar is secreted by the nectary. (This requirement for high concentration may be physiologically related as much to the correspondingly high amino acid concentration as to the sugar in the nectar (Baker & Baker, 1975).) Hence in fly flowers, rapid post-secretory evaporation would be an advantage. What features of fly flowers make this possible?

Theoretically, ideal fly flowers may be expected to produce their nectar in small aliquots (with, therefore, a relatively large surface area for evaporation): that is, there will be numerous small flowers rather than a few big ones. Their nectaries will be exposed, not set deep in a tubular corolla, and such flowers will keep their nectaries warm. For a given absolute humidity, the vapour pressure deficit of the air increases with temperature. The finding that some flowers in arctic (Kevan, 1975) and temperate (Lack, 1976) regions can focus incident radiation to produce a hot spot in the centre of the flower may be relevant here. Although these solar furnaces evidently have other advantages (Kevan, 1975), and one of the best examples, *Papaver radicatum*, is not a nectar flower at all (Hocking, 1968), it may be significant that such solar heating could speed up the daily increase in nectar concentration, and so extend the flies' visiting time. Low growing plants are also likely to heat up more quickly by day: there is commonly a steep temperature gradient near the ground in sunny weather (Geiger, 1965), so that over bare dry soil the vapour pressure deficit should be high.

These features all influence post-secretory changes in nectar concentration; but day-time concentration can also be influenced by the timing of secretion. Maximal day-time concentrations should be achieved by flowers that secrete nectar very early in the morning and then stop. Continued secretion would dilute the evaporating nectar during the heat of the day.

In contrast, in flowers adapted for visits by longer-tongued insects the concentration at which nectar is secreted is probably more nearly ideal, and over-concentration in dry conditions could reduce the frequency of visits from pollinators. What features of such flowers slow down the evaporative drying of their nectar?

Theoretically, ideal flowers adapted to long-tongued insect pollinators would produce their nectar in relatively large aliquots (with, therefore, a relatively low surface/volume ratio): that is, there would be one large flower instead of many small flowers. These would be deep flowers capable of holding a stagnant volume of humid air. As well as a long corolla tube, there may be other means of restricting exchange between the wet air inside the flower and the drier air outside: a closing mechanism, as in *Antirrhinum*; or hairs, as in *Echium*. Another possibility which has not, as far as I know, been explored, is that such flowers might secrete lipid which would coat the nectar with a monolayer of molecules, waterproofing it just as insects waterproof their cuticles (Beament, 1964). Flowers of this type should avoid over-heating, and in hot climates they may be expected to hang down instead of facing the sun; or to keep in the shade of the leaves; or to be elevated to the cooler microclimate high above the ground.

Flowers of this type may continue to secrete nectar during the activity period of the pollinator; and may well become adapted to visits from animals that are capable of flight at low ambient temperatures when the nectar is in any case dilute: at night (moths) or early or late in the day (bumblebees).

The flowers that I have described represent the extremes of adaptation to short-tongued spitters or to long-tongued tipplers. Most real flowers are likely to come somewhere in between, as *Echium* does. I have treated the pollination syndrome as if it were a fixed property of a species, adapting it for one pollinator only. But the wide fluctuations in the properties of the nectar of *Echium*, and especially *Sinapis*, show how the concentration of nectar can change through the day or from one day to another and so change the whole pollination syndrome. Some flowers may be moth flowers at night; long-tongued bee flowers in the early morning and late evening; and short-tongued bee or fly flowers in the heat of the day. The striking temporal patterning of insect visits recorded by Heinrich (1976b) on *Chamaedaphne* provides an excellent illustration of a diel progression of this sort. And it is presumably because of changes in the nectar that *Echium vulgare* receives visits from such a diversity of customers, ranging from short-tongued flies to bumblebees (and, where the corolla is short enough, honeybees) and even, in Canada, the ruby-throated hummingbird, *Archilochus colubris* (L.).

Now we can put the nectar back into its context as just one of many features of floral biology that matter to flower visitors. My excuse for this conscious over-emphasis on nectar is that it is a component of floral biology that has sometimes been undervalued in the past. To characterise the nectar of a species on the basis of a single measurement is to miss much that is of interest in a solution that changes from day to day and from hour to hour in its composition, its concentration, its viscosity and its amount. These changes reflect, and are reflected in, changes in the spectrum of flower-visitors. The dynamic complexity of nectar can be seen as just another technical obstacle to the assessment of the caloric rewards available for flower visitors; or it can be seen as a phenomenon worth studying in its own right, as a rich source of unanswered questions for zoologists and micrometeorologists as well as for botanists and beekeepers.

REFERENCES

ALLEN, G., 1891. *The Colours of Flowers as Illustrated in the British Flora.* London: Macmillan.

BAKER, H. G. & BAKER, I., 1975. Studies of nectar-constitution and pollinator-plant coevolution. In L. E. Gilbert & P. H. Raven (Eds), *Coevolution of Animals and Plants.* Austin: University of Texas Press.

BEAMENT, J. W. L., 1964. The active transport and passive movement of water in insects. *Advances in Insect Physiology 2:* 67–129.

BETTS, A. D., 1930. The ingestion of syrup by the honey bee. *Bee World, 11:* 85–90.

BOËTIUS, J., 1948. Über den Verlauf der Nektarabsonderung einiger Blütenpflanzen. *Beihefte zur Schweizerischen Bienenzeitung, 2:* 257–317.

ELTON, C. S., 1966. *The Pattern of Animal Communities.* London: Methuen.

FEINSINGER, P., 1976. Organization of a tropical guild of nectarivorous birds. *Ecological Monographs, 46:* 257–291.

GEIGER, R., 1965. *The Climate Near the Ground.* Cambridge, Mass.: Harvard Univ. Press.

HANSEN-BAY, C. M., 1976. Secretory control mechanisms in salivary glands of adult *Calliphora.* Ph.D. thesis, University of Cambridge.

HEINRICH, B., 1976a. The foraging specializations of individual bumblebees. *Ecological Monographs, 46:* 105–128.

HEINRICH, B., 1976b. Resource partitioning among some eusocial insects: bumblebees. *Ecology, 57:* 874–889.

HOCKING, B., 1968. Insect-flower associations in the high Arctic with special reference to nectar. *Oikos, 19:* 359–388.

HUBER, H., 1956. Die Abhängigkeit der Nektarsekretion von Temperatur, Luft- und Bodenfeuchtigkeit. *Planta, 48:* 47–98.

HUISH, R., 1817. *A Treatise on the Nature, Economy and Practical Management of Bees . . .* 2nd ed. London: Baldwin, Cradock & Joy.

KEVAN, P. G., 1975. Sun-tracking solar furnaces in high Arctic flowers: significance for pollination and insects. *Science 189:* 723–726.

KLEBER, E., 1935. Hat der Zeitgedächtnis der Bienen biologische Bedeutung? *Zeitschrift für vergleichende Physiologie, 22:* 221–262.

LACK, A. J., 1976. Flower-basking by insects in Britain. *Watsonia, 11:* 143–144.

LÜTTGE, U., 1962. Über die Zusammensetzung des Nektars und den Mechanismus seiner Sekretion. III. Die Rolle der Rückresorption und der spezifischen Zuckersekretion. *Planta, 59:* 175–194.

NUÑEZ, J., 1977. Nectar flow by melliferous flora and gathering flow by *Apis mellifera ligustica. Journal of Insect Physiology, 23:* 265–275.

PARK, O. W., 1929. The influence of humidity upon sugar concentration in the nectar of various plants. *Journal of Economic Entomology, 22:* 534–544.

PERCIVAL, M. S., 1965. *Floral Biology.* Oxford: Pergamon Press.

RAW, G. R., 1953. The effect on nectar secretion of removing nectar from flowers. *Bee World, 34:* 23–25.

ROBERTSON, H. A., 1974. Studies on the structure, function and innervation of the salivary glands of the moth, *Manduca sextia* (Johannson). Ph.D. thesis, University of Cambridge.

SIMPSON, J., 1964. Dilution by honeybees of solid and liquid food containing sugar. *Journal of Apicultural Research, 31:* 37–40.

S. A. CORBET

SIMPSON, J. & RIEDEL, I., 1964. Discharge and manipulation of labial gland secretion by workers of *Apis mellifera* (L.) (Hymenoptera: Apidae). *Proceedings of the Royal Entomological Society of London (A), 39:* 76–82.

STOBER, W. K., 1927. Ernahrungsphysiologische Untersuchungen an Lepidopteren. *Zeitschrift für vergleichende Physiologie, 6:* 530–565.

VANSELL, G. H., WATKINS, W. G. & BISHOP, R. K., 1942. Orange nectar and pollen in relation to bee activity. *Journal of Economic Entomology, 35:* 321–323.

WILLIAMS, G. D. V. & BROCHU, J., 1969. Vapor pressure deficit vs. relative humidity for expressing atmospheric moisture content. *Naturaliste Canadien, 96:* 621–636.

WOLF, L. L. & HAINSWORTH, F. R., 1975. Foraging efficiencies and time budgets in nectar-feeding birds. *Ecology, 56:* 117–128.

Directionality in bumblebees in relation to environmental factors

STANLEY R. J. WOODELL

Botany School, University of Oxford

Bumblebees observed during pollinating flights on *Armeria maritima* and *Limonium vulgare* at Scolt Head Island, Norfolk, showed marked overall directionality in their flight directions. This directionality appears to be related to wind direction. Its significance is discussed in relation to the direction and distance of pollen transfer.

KEY WORDS:—*Bombus*—*Armeria*—*Limonium*—Norfolk—wind—directionality.

CONTENTS

INTRODUCTION

Directionality, the tendency of successive flights of a pollinating insect to be in the same general direction; or within the same 180° category, has been surprisingly little investigated, despite the work of Bateman (1947) who showed that it existed, albeit not very strongly, in bumblebees visiting a radish crop. Levin, Kerster & Niedzlek (1971) also demonstrated it in bumblebees and butterflies visiting a population of *Lythrum salicaria*. They used 45° categories which enabled them to measure directionality with more precision. They further demonstrated that the butterflies tended to turn in one direction. Mogford (1972) pointed out that their calculations for sequential directionality ignored the fact that overall directionality was occurring. (Sequential directionality is the tendency of a flight to follow the direction of a previous flight. Overall directionality is the tendency for the majority of flights to be in one direction.) Nevertheless Mogford was able to show that bumblebees visiting wild populations of *Cirsium palustre* in Wales showed both sequential and overall directionality. All these workers have pointed out the potential importance of directionality in increasing gene flow if pollen "carry-over" occurs. None has discussed the causes for directional flights in pollinators.

During an investigation of pollination of *Armeria maritima* on Scolt Head Island, Norfolk (Eisikowitch & Woodell, 1975) I noticed that the bumblebees, the main visitors to this species, were exhibiting apparent overall directionality.

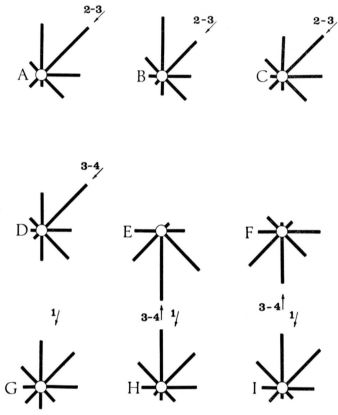

Figure 1. Bee response directions during nine foraging flights on *Limonium vulgare*. The total length of all bars represents 100% of all flower to flower flights. All flights were of *Bombus lucorum* except F which was *B. lapidarius*.

In this paper I shall give an account of subsequent observations made, in an attempt to quantify the directionality, and to determine the environmental reasons for the observed behaviour of the bees. The results will be discussed in the light of the possible effects of directionality upon gene-flow.

METHODS

Bumblebees were observed visiting *Armeria maritima* on shingle and sand-dunes in May 1975, and *Limonium vulgare* on salt-marsh, in August of the same year, both on Scolt Head Island, Norfolk. The procedure adopted was to locate a bee during a foraging flight and follow it until the flight was complete. The number of observations of flower-visits during a flight varied, depending on whether the bee was located near the beginning or end of its flight. Using a portable cassette

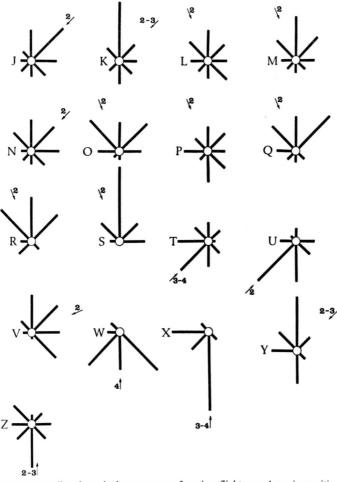

Figure 2. Bee response directions during seventeen foraging flights on *Armeria maritima*. The total length of all bars is 100% of all flower to flower flights. All flights were of *Bombus lucorum* except S which was *B. terrestris*.

recorder and a compass, the direction and duration of each flower to flower flight was recorded and later transcribed. The use of a tape recorder has the advantage that one is not distracted from watching the insect by having to write notes.

Since earlier observations had suggested that wind direction might effect directionality at this site, wind direction and speed (on the Beaufort Scale) were recorded during each foraging flight. To check whether the position of the sun was important, the time of day was also recorded, all the observations being made on clear sunny days. Observations were made in such conditions as to enable a maximum number of different wind directions to be used.

OBSERVATIONS AND RESULTS
General observations

Bees seen visiting *Limonium vulgare* were constant to that species during the period of observation. Those visiting *Armeria maritima* were usually constant, but occasionally individuals were seen to also visit *Silene maritima*. This was the only species then abundantly in flower at this site other than *Armeria*. Some of these visits were legitimate, but often the bees "robbed" the flowers of *Silene maritima* by puncturing the base of the calyx tube and corolla. Such visits may be "refuelling flights" of the type mentioned by Faegri (this Volume, p. 9), since the foraging flights may be very long.

Normally the actual direction of each flight was easy to determine; the bees flew directly from one flower to the next. Sometimes bees flying downwind were seen to turn back upwind just before alighting on flowers and these flights were recorded as downwind, even though as the individual landed they would be upwind. The phenomenon was uncommon, and was not accounted for in the calculation of results.

A bee working generally upwind would sometimes come to the edge of a patch of the plant species being visited. Often the bee was seen to take a long flight downwind and then to begin working upwind again towards the edge it had just left. All flights were recorded.

Results

Figures 1 and 2 summarise the results for nine foraging flights involving *Limonium* and seventeen on *Armeria*. They show that overall directionality

Table 1. The wind direction, speed and χ^2 calculations of differences between "into-the-wind" and "away-from-the-wind" flights during 26 separate foraging flights at Scolt Head Island. Nos 1–9 on *Limonium vulgare*, 10–26 on *Armeria maritima*. Flight 6 *Bombus lapidarius*, Flight 19 *B. terrestris*, all others *B. lucorum*.

No.	Wind direction	Wind speed	χ^2	N	P
1	NE	2–3	34.57	56	0.001
2	NE	2–3	38.25	91	0.001
3	NE	2–3	56.0	135	0.001
4	NE	3–4	29.6	95	0.001
5	S	3–4	30.1	34	0.001
6	S	3–4	87.3	212	0.001
7	NNE	1	11.2	75	0.001
8	NNE	1	38.2	88	0.001
9	NNE	1	30.2	70	0.001
10	NE	2	81.4	295	0.001
11	NE	2–3	6.8	25	0.01
12	N	2	2.4	60	0.2 NS
13	NE	2	65.3	255	0.001
14	NE	2	72.6	240	0.001
15	NW	2	10.0	29	0.01
16	NW	2	0.06	17	0.8 NS
17	NW	2	5.3	12	0.05–0.02
18	NW	2	10.3	14	0.01
19	NW	2	8.9	22	0.01
20	SW	3–4	8.5	38	0.01
21	SW	3	15.2	41	0.001
22	ENE	2	35.0	13	0.001
23	S	4	12.3	16	0.001
24	S	3–4	12.3	16	0.001
25	NE	2–3	0.25	16	0.8–0.5 NS
26	S	2–3	1.8	20	0.2–0.1 NS

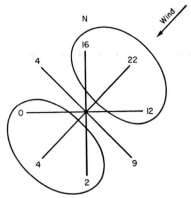

Figure 3. The division of flower to flower flights into two groups: "into-the-wind" and "away-from-the-wind".

occurred in almost all flights. The data for each foraging flight was treated as two blocks, of "into-the-wind" and "away-from-the-wind" flights, ignoring those flights at 90° from the wind direction (Fig. 3) and chi-square calculations showed that in most flights there was a significant majority of "into-the-wind" flights. Those which did not give significant results were foraging flights in which few visits were recorded (Table 1).

Despite Mogford's comments (1972) that where overall directionality exists, sequential directionality will be exaggerated, calculations of sequential directionality were made, using the method of Levin *et al.* (1971) and in most flights

Table 2. Sequential directionality for 26 foraging
flights at Scolt Head Island, Norfolk

No. of flight	Sequential directionality
1	0.605
2	0.499
3	0.507
4	0.576
5	0.612
6	0.377
7	0.397
8	0.481
9	0.593
10	0.512
11	0.383
12	0.506
13	0.466
14	0.586
15	0.358
16	0.207
17	0.520
18	0.327
19	0.620
20	0.376
21	0.300
22	0.498
23	0.622
24	0.677
25	0.180
26	0.622

strong sequential directionality was shown (Table 2). The justification for making these calculations lies in the fact that regardless of overall directionality, there is a very strong tendency for a flight to be in the same or similar direction to the previous one. This is indicated in Fig. 4, where the frequency distribution of flight directions subsequent to each base flight direction (response flights) indicates that the frequency of these secondary flights is at a maximum in the direction of the base flight in every direction except west. Thus quite apart from overall directionality, sequential directionality is very consistent, i.e. the direction of any one flight is strongly influenced by that of its predecessor.

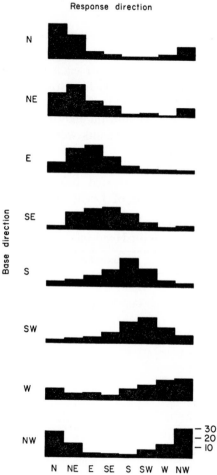

Figure 4. Directionality in successive flights, Scolt Head Island, June and August 1975. Vertical axes represent percentage frequency distributions of response flights.

Knowing the total time spent on one foraging flight, and the number of flower visits made, the mean time spent per flower flight-visit can be calculated. When visiting *Armeria maritima* the mean time/visit was 4.96 ±0.37 sec. For *Limonium vulgare* the mean time/visit was 6.87 ±0.59 sec. *Limonium* inflorescences are many-flowered and bees may spend longer on each one.

DISCUSSION
The causes of directionality

The observations made here are in agreement with those of Levin *et al.* (1971) and Mogford (1972). Bumblebees of three species all showed strongly directional trends in their foraging flights. Levin *et al.* (1971) made no mention of the possibility that environmental factors may control directionality; they pointed out that in their study, the tendency for pollinators to move in the same general direction related only to successive flights and did not yield strongly directional feeding forays. It is not clear whether this absence of overall directionality was the result of the wind speed being too low to affect the insects (winds were described as "light and variable") or whether the insects were not responding to wind in the same way as those at Scolt Head Island. They suggested that directionality might be a general adaptation to reduce the probability of returning to the same plant soon after it was visited.

Mogford (1972) found that in south Wales, bees visiting *Cirsium palustre* showed both sequential and overall directionality, like those at Scolt Head Island, and he suggested that wind might be responsible. Winds were light during the period of his observations and he measured neither their strength nor their direction. However, the fact that on two successive days he found that overall directionality showed significant bias in different directions suggests that wind might be the causal factor. He commented that it was difficult to see how strong winds could fail to produce overall directionality, though he did not indicate whether it would be upwind or downwind.

My observations demonstrate that overall directionality can exist and that wind is almost certainly the causal factor. The reversal of this directionality with a change in wind direction is particularly striking. The response of honeybees to wind direction is well known, but it has been discussed mainly in relation to flights from the hive to the foraging grounds, not in flower to flower flights. The reason for the bees' flights being largely upwind could be either aerodynamic, or because they are responding to odour from flowers, or possibly in a nodding inflorescence like *Armeria*, because it is more visible to a bee flying upwind. The last possibility could not hold for *Limonium vulgare*, however. *Armeria* pollen has a strong fragrance in large quantity, and it is possible that bees are flying up an odour stream. The fact that bees are not always flying *directly* upwind, but are often flying at an angle to the wind though in a general upwind course, would seem to rule out this view. The most likely explanation, in my opinion, is that bumblebees find it much easier to land on a small, and often moving inflorescence, when flying into the wind. I have observed bees being carried past flowers when flying downwind, unable to gain a foothold as the wind carries them along, but flying into the wind they can regulate the speed at which they land, and have little difficulty in doing so. The occasional reversal from downwind to upwind just before landing appears to be one method of achieving this regulation.

The consequence of directionality

The importance of directionality in plant population studies lies both in the fact that it may prevent flowers being revisited too soon, as Levin *et al.* (1971) suggested, and in the possible results of pollen "carry-over". Bumblebees carry considerable amounts of pollen on their bodies; though they may frequently stop and sweep the pollen into their corbiculae, this is not done between every flower and some pollen "carry-over" occurs. Gerwitz & Faulkner (1972) showed, by labelling pollen with a radioactive isotope, that the amount of radioactivity detectable in a sequence of flowers visited by honeybees (*Apis mellifera*) decreased by an average of 30% at each of the flowers and labelled pollen was still detectable at the tenth flower. If flight directions are random then the effect of "carry-over" would be relatively unimportant. Where directionality, is operating, then "carry-over" will be highly important in determining the distance of pollen transfer, and hence gene-flow, in natural populations.

In most places winds are variable in direction and do not blow in one direction for long periods. However, in a coastal locality, there tends to be a daily pattern of morning offshore and afternoon onshore breezes; thus the effects of overall directionality will cancel each other out. Nevertheless, the relatively strong winds encountered at the coast are likely to encourage directionality at any one time and thus increase pollen transfer distances, perhaps especially towards and away from the coast-line.

If winds persist from one quarter, the direction of pollen transfer, as well as its distance is likely to be affected. In many tropical areas the wind pattern is one of prevailing trade winds during a large part of the year; in the western Indian Ocean for instance, the winds are in the south-east for several months every year. In places like this unidirectional pollen flow may be very important, especially if a species' flowering period coincides with the winds of one direction. Even in temperate areas there are localities where winds are overwhelmingly unidirectional. A good example of how such winds may affect pollen flow is that described by McNeilly (1968) who showed that populations of *Agrostis tenuis* in the vicinity of a small isolated copper mine in north Wales which was subject to strongly unidirectional winds, changed suddenly from predominantly tolerant to predominantly non-tolerant individuals over a distance of one metre upwind of the mine, but downwind populations 150 m away from the mine boundary had a high proportion of tolerant individuals. *A. tenuis* is wind-pollinated, but it is not difficult to envisage insect-pollinated plants behaving similarly in such an area, if the insects behave in the same way as the bees at Scolt Head Island. In areas with steep environmental gradients, e.g., from a coast inland, unidirectional gene flow towards the coast could maintain a higher degree of tolerance coastwards without dispersing non-adaptive salt-tolerant genes inland.

ACKNOWLEDGEMENTS

Thanks are due to the Nature Conservancy Council, who gave me permission to work on Scolt Head Island. C. O'Toole identified the bees for me. F. Topliffe drew the figures. Valuable comments on the paper have been made by Dr. D. Eisikowitch.

REFERENCES

BATEMAN, A. J., 1974. Contamination n seed crops. III. Relation with isolation distance. *Heredity, 1:* 303–336.
EISIKOWITCH, D. and WOODELL, S. R. J., 1975. Some aspects of pollination ecology of *Armeria maritima* (Mill.) Willd. in Britain. *New Phytologist, 74:* 307–322.

FAEGRI, K., 1978. Trends in research on pollination biology. In A. J. Richards (Ed.), *The Pollination of Flowers by Insects:* 5–12. London: Academic Press.

GERWITZ, A. and FAULKNER, G. J., 1972. *Report of the National Vegetable Research Station for 1971:* 32. Wellesbourne.

LEVIN, D. A., KERSTER, H. W. & NIEDZLEK, M., 1971. Pollinator flight directionality and its effect on pollen flow. *Evolution, 25:* 113–118.

MCNEILLY, T., 1968. Evolution in closely adjacent plant populations. III. *Agrostis tenuis* on a small copper mine. *Heredity, 23:* 99–108.

MOGFORD, D., 1972. The ecological genetics of flower colour variation in *Cirsium palustre*. D.Phil. thesis, University of Oxford.

The possible role of insect visits in pollination of reputedly anemophilous plants, exemplified by *Plantago lanceolata*, and syrphid flies

P. STELLEMAN

Hugo de Vries Laboratorium, University of Amsterdam, The Netherlands

Frequent records of visits by certain Syrphidae to *Plantago* and other plants exhibiting a nominal anemophilous syndrome, combined with diet studies of these hover flies, raise the question whether such visits may lead to an effective pollen transfer by these insects. Examination of the flies by means of the SEM technique shows that normal and artificially stained pollen grains of *Plantago* adhere long enough to the fly body to become transported to other plants in anthesis. By means of dyed pollen grains applied to flowering spikes of *Plantago*, the uptake of pollen by flies and the subsequent deposition on a receptive stigma of a different specimen of the plantain was demonstrated experimentally. The conclusion is drawn that such insects may indeed be important vectors of *Plantago lanceolata* pollen and conceivably contribute towards the effective pollination of this plant.

KEY WORDS:—*Plantago*—Syrphids—anemophily—pollen marking—SEM.

CONTENTS

INTRODUCTION

Among the herbaceous angiosperms of the temperate regions there are several groups mainly consisting of anemophiles. Examples of such taxa include larger assemblies (such as Gramineae and Cyperaceae) and smaller ones (such as Plantaginaceae, Urticaceae, and Juncaceae). Most of these exhibit a number of characteristics indicating a more or less complete adaptation to anemophily, so that their predominantly anemophilous nature is fairly clear. It is, therefore, rather striking that reports have continually appeared concerning visits of pollen-consuming and pollen-gathering insects to flowers or inflorescences of such plants with an apparently anemophilous syndrome. Müller (1873), Knuth (1899) and Porsch (1956), to mention only some of the better known authors, have mentioned such visits. Some observers also believe that the visiting insects conceivably play a role in pollen transfer in addition to wind pollination. However, one must not lose sight of the fundamental issue of an effective pollination, because suppositions to that effect were almost entirely based on records of visits. Since the observations in the field were mostly incidental, such data by themselves do not warrant any definite conclusions regarding a possibly consistent contribution from the insects.

A study by van der Goot & Grabandt (1970) is important in this connection. By studying the contents of the digestive tract of a number of species of hover-flies they found that the diet of several of them consisted predominantly or exclusively of pollen of a plantain (*Plantago lanceolata*) and/or of grasses (or sedges: Grabandt, unpubl.).

The scattered records published in the anthecological literature and the results of the above-mentioned diet studies appeared to us to be important enough to investigate the subject more extensively. It was thought that the primary issues can be expressed in the following two questions:

(1) are the syrphid flies under discussion indeed regular visitors of anemophiles?

(2) and if this appears to be the case, are they capable of an effective pollen transfer?

The studies by van der Goot and Grabandt indicated that (in the Netherlands at least) four species of Syrphidae come into consideration as potential pollen-vectors, *Melanostoma mellinum* (L.), *Platycheirus clypeatus* (Meigen), *P. fulvi-ventris* Macquart, and *P. scambus* (Staeger). When in this paper "syrphids" are mentioned, these four species are meant as an ecological group. Furthermore, we shall restrict ourselves to the protogynous anemophile *Plantago lanceolata*, because so far the anthecological relations between this plant species and the syrphids in question have been studied most extensively. It may be mentioned in passing that the results of preliminary studies indicate that the conditions are probably very much the same in *Plantago coronopus* and *P. maritima* but rather different in *P. media*. It is also necessary to point out that suitable habitats for such studies are low-lying sites near water (the species *Melanostoma mellinum* is not so particular, however: we found it in many dry localities, even on arid limestone slopes near Steinau in eastern Hessia).

Three methods of approach were employed:

(1) visual observations in the field with the purpose of establishing some overall relationship between *Plantago* and syrphids;

(2) examination of syrphids by means of stereoscan apparatus, which provides a unique possibility to locate pollen grains attached to the insect body (one must ascertain whether pollen sticks to the insects for some time and is not removed by preening, etc.);

and (3) it was attempted by means of artificially dyed pollen, to demonstrate experimentally a transfer of pollen by flies from one plant to another.

The results obtained by the use of these three methods will be briefly reported here.

METHODS

Observations in the field

The results of a large number of observations in various sites in the Netherlands and in West Germany are in so far consistent that we may safely conclude that the syrphids under study are habitual consumers of plantain pollen. The visits are usually not very extensive but nevertheless clearly directional. As an illustration: in the Naardermeer nature conservancy in the Netherlands the mean number of flowering spikes visited at least once during a number of days with varying weather conditions was as high as 80%. I am of the opinion that the first question can be answered in a positive sense with the possible restriction that this only holds for the areas where our field studies were carried out.

A second interesting result is that there is a striking parallel between the daily commencement of the male phase of anthesis and the beginning of the activities of the flies. Both phenomena start in the early hours of the morning, usually about the time of sunrise. There is a prerequisite in that the temperature must not be too low and must be between 10° and 13°C, dependent on the other weather conditions (rain or mist or sunshine). In terms of solar time the events take place between 05.00 and 07.00 hrs. After a fairly slow start the intensity of visiting increases rapidly in a short time, but the period of optimal feeding is relatively short and only rarely lasts for more than a few hours, to cease by about mid-morning; subsequent visits are clearly incidental, most of the animals start a period of rest. They are then manifestly less active, which is evident from the fact that they land on all kinds of plants and various parts of these plants but do not ingest pollen even if they sit on an inflorescence, though preening often takes place. Unfavourable weather conditions—more particularly rain or drizzle during the night or early morning—may alter the daily rhythm and cause a shift in the visiting hours.

Some behaviour patterns of the syrphids may be of direct importance for a possible pollen transfer, particularly their mode of descent and movement on the flowering spikes of *Plantago lanceolata*. When a fly decides to visit an inflorescence it mostly descends inside the whorl of stamens and immediately starts ingesting pollen, its body coming into contact with the anthers, but not rarely a fly first lands on the part of the spike above the stamens, i.e. on the part where only the female phase of anthesis has occurred. From this position on the spike, the fly walks downwards towards the stamen-bearing zone, and in this way its body may brush past the mature stigmas, but there are also other possibilities of a direct contact between the fly and the stigmatic surfaces (such as a landing in the middle of the zone of female anthesis).

These observations show unequivocally that a number of essential prerequisites for an effective pollen transfer have been fulfilled. We subsequently had to ascertain whether pollen grains get stuck on the fly body and can thus be transported across certain distances. This formed the second part of our study by means of stereoscanning (SEM).

SEM studies

The pictures obtained by means of stereoscan electron microscopy provide an ideal method both for the study of the localisation of the pollen grains on the insect body and for the specific identification of the attached grains. A number of syrphids was collected, sometimes immediately after a visit to a flowering specimen of *Plantago*. After mounting, the specimens were studied by means of the electroscan in order to ascertain the presence of pollen grains. All grains present on a specimen were drawn in the approximate position on a semi-diagrammatic drawing of the insect in order to obtain some statistical data. Some characteristic positions were recorded on photographs. The results can be briefly summarised as follows: on the ventral side of practically every specimen examined, pollen grains were found. These were almost invariably grains of *Plantago lanceolata*; gramineous pollen was only encountered occasionally, and other species were totally absent. The number of grains adhering to a single specimen varied from one to 179 (with a mean value of 44). Most of these grains were located on the thorax, especially on the pleural parts and in the zones of hairs present in these parts. However, a well-founded conclusion concerning this distribution pattern on the fly body would be premature in view of the relatively small number of specimens studied. Pollen grains are found in places most likely to come into contact with a stigma. When assessing the results of these records of attached grains one must bear in mind that during the capture and killing of the insects, and also during the mounting for the SEM some of the grains may probably be rubbed off, so that the impression one gets may be biased.

In combination with our field studies the SEM records provided a very strong indication of a biotic pollen transfer. One link in the chain was still missing, however, namely evidence of the deposition of insect-borne pollen on mature stigmas. To establish the probability, appropriate experiments in the field were carried out.

Experiments in the field

This could be met by working out a method in which the use of artificially stained pollen of *Plantago* (indicated here as S-pollen) forms an essential element. It was shown that S-grains can act as easily recognisable replacements of normal, unstained grains, but no details will be given here. It will be clear, however, that it thus becomes possible to trace the pathway of insect-borne grains under field conditions. This can be accomplished by means of an experimental set-up as shown in Fig. 1. Two groups of flowering spikes are selected and to one group S-pollen is applied as explained below. We shall call these the "donor spikes". The other group of spikes is left intact and they act as "receptor spikes". We start from the supposition that a fly visiting a donor spike can pick up some S-pollen and, if it subsequently lands upon a receptor spike, may deposit S-grains on a stigma. After the experiment this can be confirmed by examining the receptor spikes under a low-power microscope. The factor of air currents must of course be taken into account: S-pollen may also become airborne. In order to reduce contamination to a bare minimum the donor spikes were always downwind of the receptor spikes. A favourable circumstance is that the hover flies during their foraging flights fly against the direction of the wind as a rule, in this case from donor spikes to receptor spikes. It is clear that this typical behaviour enhances the efficacy of the experiment. As a special precaution, a control measure was taken by placing sticky test slides (facing the donor spikes with their adhesive surface) between the donor spikes and the receptor spikes so as to catch any S-grains that might have been abiotically transported through the air by convection currents or otherwise.

The S-pollen was produced by spraying freshly harvested pollen with an aqueous solution of methylene blue or neutral red in a very simple way. An

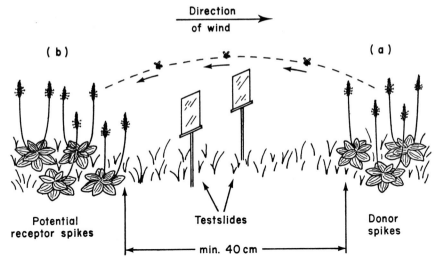

Figure 1. Situation sketch of a field experiment (explanation in text). For the sake of clarity the rosette of leaves is drawn as if they are adpressed to the ground (as in *Plantago major* and *P. media*).

important consideration was that the sticking power of dyed grains should remain approximately the same as that of untreated pollen. This was put to the test in a small experiment designed for the purpose which showed that normal pollen grains adhere slightly better to a fly body than S-grains (for the purpose of the field experiments the small difference is of course unimportant). The S-pollen is applied to a spike as follows: A small quantity of S-pollen is put in a glass tube about 2 cm in diameter and divided across the inside of the tube as evenly as possible by shaking and tapping. In the field a spike with well-developed stamen whorl is selected, bent and stuck into the tube in a horizontal position after which it is moved to and fro several times so that the S-grains may adhere to the inflorescence, especially on the protruding stamens (Fig. 2). The stamens may assume a bluish or reddish colour, but this does not seem to have any deterrent effect upon the flies. S-pollen is not only carried passively on the fly body but is as readily consumed as unstained pollen.

For additional details the reader is referred to Stelleman & Meeuse (1976) and to Leereveld, Meeuse & Stelleman (1976).

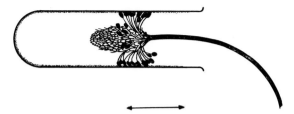

Figure 2. Application of stained pollen to a flowering spike of *Plantago*. A short glass tube of suitable diameter is covered on the inside with stained pollen, placed over the spike which is bent into a horizontal position, and pushed to and fro in the direction of the double-pointed arrow.

RESULTS AND DISCUSSION

During the summer of 1974, 22 complete experiments were performed, which were augmented by an additional eight the next year. After every experiment the test slides were perused microscopically; since not even a single S-grain was found, contamination with wind-borne S-pollen must be so negligible that as a disturbing factor it can altogether be ruled out.

The receptor spikes which had been visited by flies presumably carrying S-pollen were carefully collected and examined for the presence of stained pollen grains. Three interesting points emerged:

    (1) after each experiment S-pollen was found to be present on a number of receptor spikes, and

    (2) of the total of these receptor spikes, 64% appeared to have at least one receptive stigma on which S-grains were attached.

    (3) on an average of 30% of the receptive stigmas of receptor spikes, S-grains were recorded (i.e. often after only a single visit).

These results warrant the conclusion that the syrphids can carry S-pollen from one inflorescence to another one and in fact do this fairly regularly. It follows that they can act as vectors of untreated pollen at least as effectively. We must also take into account that the syrphids in the experiments also carried normal pollen grains from the donor to the receptor spike. This we cannot assess quantitatively however, because the normal grains we observed on the same stigmatic surface as the S-grains may have been transferred by air currents or by different (previous) visits of flies. We may nevertheless accept that the total rate of biotic pollen transfer indubitably exceeds the amount computable from the experimental results obtained.

Our future experimental studies will especially deal with the question in how far insect pollination contributes appreciably towards the setting of seed.

REFERENCES

GOOT, V. S. VAN DER & GRABANDT, R. A. J., 1970. Some species of the genera *Melanostoma, Platycheirus* and *Pyrophaena* (Diptera, Syrphidae) and their relation to flowers. *Entomologische berichten, 30:* 135–143.

KNUTH, P., 1898–1905. *Handbuch der Blütenbiologie, I–III.* Leipzig.

LEEREVELD, H., MEEUSE, A. D. J. & STELLEMAN, P., 1976. Anthecological relations between reputedly anemophilous flowers and syrphid flies. II. *Plantago media* L. *Acta botanica neerlandica, 25:* 205–211.

MÜLLER, H., 1873. *Die Befruchtung der Blumen durch Insekten.* Leipzig.

PORSCH, O., 1956. Windpollen und Blumeninsekt. *Oesterreichische botanische Zeitschrift, 103:* 1–18.

STELLEMAN, P. & MEEUSE, A. D. J., 1976. Anthecological relations between reputedly anemophilous flowers and syrphid flies. I. The possible role of syrphid flies as pollinators of *Plantago. Tijdschrift voor entomologie, 119:* 15–31.

Entomophily in *Salix*: theoretical considerations

A. D. J. MEEUSE

Hugo de Vries Laboratorium, University of Amsterdam, The Netherlands

The frequency of insect visits recorded in *Salix* and in other diclinous taxa does not warrant the conclusion that such visits contribute appreciably to the specific transfer of pollen from male polliniferous organs to the stigmas of their female counterparts. Such a translocation of pollen must be established beyond reasonable doubt before any assessment can be made of the efficacy of the visits and thus of the possible role of zoophily in a given diclinous taxon.

KEY WORDS:—*Salix*—diclinous—entomophilous—anemophilous.

Frequent visits to *Salix* catkins by insects has led to the belief that this genus is an example of an "intermediate" stage between a supposedly more primitive, entomophilous syndrome, and the allegedly derived anemophilous one. Examples of the various expressions of this viewpoint can, *inter alia*, be found in Percival (1965: 60) "*Salix* . . ., despite its looks, it is truly entomophilous", and in Faegri & van der Pijl (1971: 160): "The idea of revertence (i.e. reversion to anemophily, A.M.) only holds if one projects backwards beyond the origin of angiospermy to hypothetical ancestors, . . ." (which means that by implication, the early Angiosperms—including early Salicales forms—were entomophilous). For some other details, additional references, and some relevant criticism, see Meeuse (1972b, 1973).

One of the frequently adduced arguments said to support the opinion the *Salix* is "on its way to anemophily" is the idea that it has "retained" nectaries, a condition supposed to have persisted from earlier "proto-*Salix*"; compare B. J. D. Meeuse (1961: 74) "The answer is that they (nectaries in *Salix*, A.M.) did not have to be conjured up *because they had been available all the time*" (italics mine).

It is quite clear that two aspects of the anthecological conditions in willows are most relevant in the present context, the floral morphology on the one hand, and the efficacy of animal pollination, in comparison to pollen transfer by air currents, on the other. The floral morphology has been discussed in great detail in a previous publication (Meeuse, 1975), mainly as regards its probable bearing on the phylogeny and taxonomic position of the Salicaceae. The question of the significance of hermaphroditism in *Salix* has been reduced to two points of inquiry: (1) are the hermaphrodite catkins and florets more primitive than the unisexual ones? and (2) is this anomalous kind of sex distribution interpretable as an atavistic recurrence of a one-time common (and normally hermaphrodite) phenotype; or merely as a teratological aberration, which has no importance as regards the primitiveness of monocliny (and hence, of entomophily) in early taxa in the Salicales?

The answers are that: (1) hermaphrodite florets are not necessarily primitive on the grounds of comparative morphological and taxonomic criteria, and (2) their occasional and irregular occurrence—for instance in unseasonal summer catkins— is not an atavistic phenomenon, but simply an anomaly which, if it has any morphological meaning at all, would rather plead in favour of a progressive change-over from dicliny to monocliny, than of a retrograde return ("throw-back") to a primarily hermaphrodite condition associated with phaneranthy (i.e., with entomophily and petalifery).

Another morphological question arising, if the primitiveness of anemophily in *Salix* is accepted, is the origin of the nectaries in *Salix* florets. As stated already, the suggestion has repeatedly been made that the presence of nectaries is indicative of an erstwhile entomophilous floral syndrome, but this idea cannot be maintained in the light of morphological and taxonomic evidence. Hjelmqvist (1948) suggested that the nectaries are derived from some perianth- or disc-like structure present in a few Asian species of *Salix*, which must, therefore, be completely wind-pollinated. As far as can be ascertained, the absence of nectaries in pistillate florets would preclude visits by insects because of the lack of any attraction—there is nothing of interest offered to prospective pollen vectors.

Alternative interpretations of the nectaries as modified bracteoles (or even stamens!) do not alter the *secondary* nature of the nectar-secreting organs. The primitiveness of pistillate florets without nectaries, consequently, not only renders the primitiveness of entomophily but also, implicitly, of monocliny and petalifery in *Salix* most unlikely.

The efficacy of the visits of pollinators in willows or sallows has not been assessed, but one requisite is quite obvious: to be successful as pollen vectors, visiting insects must be instrumental in the transportation of pollen from a male plant to a female specimen of the same species of *Salix*. We can immediately rule out all visitors ingesting pollen alone (such as most beetles), because they would not find anything attractive in the female florets and would restrict their visits to male catkins exclusively. However, it has repeatedly been established that *Salix* is frequently visited by honey bees (*Apis*) (Plate 8), bumblebees (*Bombus*) and other apiid Hymenoptera (such as species of *Andrena*), insects which gather both pollen and nectar (mostly at the same time when both are available). Among other prospective pollen vectors, flower-visiting Diptera such as hover flies (Syrphidae), although not so constant in their food plants, also collect both kinds of food as a rule. According to Van der Goot (pers. comm.) the syrphid *Melangyna quadrimaculata* Verrall is a typical *Salix* visitor as early as March.

The problem is even more complicated because honey bees, and possibly other social Hymenoptera (bumblebees, etc.) may also be attracted by specific flower scents, sometimes in a special way (e.g. after instruction by honey scouts: *Apis*). The question arises in how far these insects exhibit flower constancy: conceivably, they may come across a male specimen of *Salix* on their first gathering flight and remain faithful to the polliniferous kind of plant (or possibly to the same speci-men!), or they may encounter a female willow and develop a constancy, after the first visual contact, for the pistilliferous specimens or for a single female plant, as the case may be. We simply do not know in how far instruction and experience play a part and can only guess. However, it is by no means a foregone conclusion that the Hymenoptera visiting *Salix* consistently "alternate" between male and female specimens, which is a prerequisite for a high rate of effective pollination. It is evident that we must study the behaviour of visiting insects, for instance by marking them individually and by recording their visits in a suitable locality (e.g. in an isolated group of willows or sallows). Only after having established the flight pattern can we hope to assess the relative efficacy of the visits and the degree of dependence on wind-borne pollen transfer.

The problem is not restricted to *Salix*. The primitiveness of phanernathy and entomophily has been accepted for such currently predominantly anemophilous groups as Amentiferae, Hamamelidales, Urticales, Cyperales, and Palmae. The same objections can be raised as in the case of the willow family: how effective are the visits by insects in reality? Regular visits by insects to palm inflorescences do not prove anything as regards an effective entomophily until a specific pollen transfer has been demonstrated beyond reasonable doubt (compare the case of *Plantago*: Stelleman & Meeuse, 1976; for theoretical considerations, see Meeuse, 1972a). The same can be said about the Hamamelidales, Cyperales, Urticales, and the amentiferous orders. The present author (Meeuse, 1975: 168–170) has adduced arguments indicating a highly primitive androecial morphology in Juglandaceae (which would refute an advanced status of the family), and in a forthcoming paper (Meeuse, in press) the idea of a conventionally advanced and "reduced" floral morphology in the Platanaceae has been shown to be untenable: the primitiveness of entomophily and phaneranthy in the hamamelidid group is, therefore, highly questionable to say the least. The attractiveness of the male florets of *Castanea* to insects (see Faegri & van der Pijl, 1971: 159) is another misinterpreted phenomenon. It is immaterial whether numerous individuals of insects, even if they belong to scores of species, are enticed to visit its polliniferous florets as a source of food as long as the female catkins remain uninteresting and are completely ignored.

A very similar case has been discussed by Frankie, Opler & Bawa (1976): self-incompatible species depend on outcrossing for their reproduction, so that entomophily can only be efficacious if available pollen vectors exhibit inter-tree visiting. Even if a tree specimen flowers profusely, seed-setting can only come about by movement of the insects between conspecific trees (provided wind-pollination can altogether be excluded). The problem is somewhat different from the "true" (not functional) dioecy in *Salix*, in that self-incompatibility is associated with ambisexuality (monocliny or monoecy). Nevertheless, in both cases the solution lies mainly in a study of marked specimens of the insect visitors in order to establish the rate of insect traffic between conspecific individuals.

Summarising, it is necessary to realise that the registration of insect visits to diclinous and presumably—at least facultatively—anemophilous flowering plants has no bearing on the principal issue whether entomophily *is effective* (i.e. whether or not such forms depend to a large extent on animal pollen vectors for their reproduction). Only after having established "cross-visits", an alternation of visits to male and to female flowers or florets (or in many cases: between male and female plants), can we assess the possible role of insect pollination in diclinous Angiosperms.

A. D. J. MEEUSE

REFERENCES

FAEGRI, K. & van der PIJL, L., 1971. *The Principles of Pollination Ecology*, 2nd ed. Oxford: Pergamon Press.

FRANKIE, G. W., OPLER, P. A. & BAWA, K. S., 1976. Foraging behaviour of solitary bees: implications for outcrossing of a neotropical forest free species. *Journal of Ecology, 64:* 1049–1057.

HJELMQVIST, H., 1948. Studies on the floral morphology and phylogeny of the Amentiferae. *Botaniska notiser (Suppl.), 2*(1): 1–171.

MEEUSE, A. D. J., 1972a. Palm and pandan pollination—primary anemophily or primary entomophily? *Botanique (Nagpur), 3:* 1–6.

MEEUSE, A. D. J., 1972b. Angiosperm phylogeny, floral morphology and pollination ecology. *Acta biotheoretica, 21:* 145–166.

MEEUSE, A. D. J., 1973. Anthecology and Angiosperm evolution. In V. H. Heywood (Ed.), *Taxonomy and Ecology:* 189–200. London: Academic Press.

MEEUSE, A. D. J., 1975. Floral evolution in the Hamamelididae. II. Interpretative floral morphology of the Amentiferae. *Acta botanica neerlandica, 24:* 165–179.

MEEUSE, A. D. J., 1976. Taxonomic relationships of the Salicaceae and Flacourtiaceae: their bearing on interpretative floral morphology and dilleniid phylogeny. *Acta botanica neerlandica, 24:* 437–457.

MEEUSE, B. J. D., 1961. *The Story of Pollination*. New York: Ronald Press.

PERCIVAL, M. S., 1965. *Floral Biology*. Oxford: Pergamon.

STELLEMAN, P. & MEEUSE, A. D. J., 1976. Anthecological relations between reputedly anemophilous flowers and syrphid flies. I. The possible role of syrphid flies as pollinators of *Plantago. Tijdschrift voor Entomologie, 119:* 15–31.

Floral coloration, its colorimetric analysis and significance in anthecology

P. G. KEVAN

Biology Department, University of Colorado, Colorado Springs, U.S.A.

Floral colours have long fascinated students of pollination. The discovery of ultraviolet as an insect colour added a further dimension to analysis of floral allurements and to the complexities of anthecology. Unfortunately, ultraviolet patterns have received almost all the attention, whereas other colours, which are also an integral part of daylight and colour vision, remained all but ignored. To fully understand the role of floral (and other) colours in attracting insects, full colorimetric analyses must be made in relation to the colour vision systems of insects and to the ambient colour spectrum in which they live. Following from a physical and universal definition of white, a colorimetry scheme can be erected to define floral colour with respect to the daylight spectrum and insect colour vision. The proposed scheme relies heavily on published studies on honeybee colour vision, but has wider applicability.

The application of this scheme in anthecology is discussed especially as it relates to pollinator recognition of flowers and floral structures, competitive interactions, phenology, and pollination ecology generally. Examples from the High Arctic, the Colorado Alpine, and weedy habitats near Ottawa, Canada, are discussed.

KEY WORDS:—floral colours—ultraviolet—definition of white—equal energy white point—insect vision—weeds—alpines.

CONTENTS

INTRODUCTION

The spectrum of solar radiation on the surface of the earth is broad, extending from about 290 nm to 3000 nm in favourable conditions at sea level. Vision in all animals is restricted to the shorter wavelengths, from about 290 nm to 800 nm. This paper examines the visual spectrum of insects (IVS) and the measurement and definition of colours therein, especially as these colours relate to the attractiveness of flowers to insect pollinators (Fig. 1). It is convenient to draw from the works on

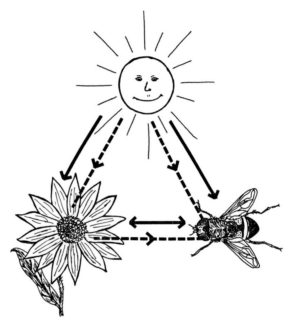

Figure 1. Daylight and anthecology. Dashed lines show the paths of light. Light emitted from the sun and sky forms the photic environment for both blossoms and anthophiles. Light is reflected from the blossom to the eyes of anthophiles, in which coloured images of the blossoms may act as attractants for potential pollinators. Hence a co-adaptive scheme (double ended solid arrow) may result. Single ended solid arrows show that both blossom and anthophile are adapted to daylight, the blossom in its reflectance and the anthophile in its spectral sensitivity.

human colour vision and the spectrum of human visual sensitivity (HVS, about 380 nm to 780 nm). IVS and HVS together constitute a good representation of the animal visual spectrum (AVS).

I will examine briefly the physical aspects of daylight, especially in the AVS, before discussing colour vision and introducing some concepts important to the measurement of colour, particularly in the IVS. From there, the importance of floral colours will be considered.

Daylight

Henderson (1970) has reviewed the subject of solar and daylight spectra. He summarises the difficulties in defining daylight as follows:

> "The term 'average daylight' has been used frequently in the past, but with little justification if applied to experimental measurements of the spectrum. Sampling has rarely extended systematically over the whole of the daylight hours and never in this way for a long period. The matter is raised because daylight is used physiologically, with spectral discrimination, by land plants and many animals wherever available, much of it outside man's active hours . . . Though there is plenty of evidence of the total power and total light received at different places and seasons, there is none of spectral power distribution. . . . So far determinations of the mean have scarcely progressed beyond the collection of many spectral curves on different days and at different times, usually with normalization of the curves at a selected wavelength . . .". (pp. 187–188).

It is from this "collection of many spectral curves" that generalisations must be made.

For colour measurement, it suffices within limits to rely on the relative amounts of energy in daylight (normalised curves). Much of the solar energy (about 65%) reaching the surface of the earth is in the AVS (Fig. 2) and the relative amounts of energy in wavebands in the AVS are more constant than in the infrared (IR) which is greatly affected by absorbtion by water vapour, CO_2, aerosols, dust, etc., which vary widely spatially and temporally.

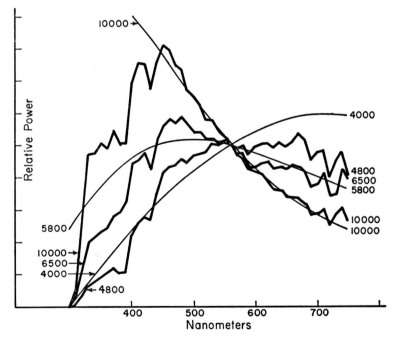

Figure 2. Relative amounts of power in daylight for correlated colour temperatures of 4800°K (from Judd *et al.*, 1964), 6500°K (from Henderson & Hodgkiss, 1963), and 10,000°K (from Judd *et al.*, 1964) (solid lines) compared with a Plankian radiator at 4000°, 5800°, and 10,000°K.

The distribution of energy in the AVS depends on several atmospheric factors, air mass (M) (including solar elevation and altitude ASL), and turbidity and cloudiness. Complications arise as solar radiation reaches the earth's surface directly (D) or scattered from the sky (S). Together S and D constitute global radiation (G) which is defined as incident on a horizontal plane. S radiation also varies depending on aspect. For example, in the Northern Hemisphere, north sky light has a range of correlated colour temperatures (see below) from well above 11,000° to 5000°K, and south sky light from 3000° to 4800°K. Hence, the orientation of a coloured object can also affect its apparent colour, as will become evident. Increasing M increases the amount of energy in S at the longer wavelengths (i.e. from 500 nm on), accounting for the reddish hues of sunset and apparent changes in colours then.

Colour temperature

As an object is heated, it first retains its original colour, then begins to emit radiation at the red part of the HVS. At that and subsequent points, the object's temperature can be measured with its colour. Colour temperature is, then, the temperature of a full radiator which emits radiation matching (more or less) the colour, or chromaticity, of the source considered (Harding, 1950). When describing or measuring the colour of daylight, the correlated colour temperature (CCT) is frequently used. The Plankian relationship provides a formal baseline for calculating the amount of energy at any wave length being emitted from a fully radiating body at any temperature.

Generally the CCT of daylight lies between 5500 and 6200°K; 5800°K can be considered a useful standard. Figure 2 shows the energy distribution of a Plankian radiator at 10,000°, 5800° and 4000°K, and of some daylights at various CCT.

Colour measurements of daylight in the HVS

The measurement of colour and the analysis of daylight has been a subject of systematic study since the time of Newton (1704) and earlier (*vide* Henderson, 1970).

All colours perceived by humans can be described in terms of three primary colours (Young-Helmholtz) (MacNichol, 1964). This provides the foundation of colour measurement in the trichromatic system, wherein the proportion of each primary colour (wavelength) reflected or emitted by an object can be used to plot, on a trichromaticity triangle, the colour of the object. Several textbooks explain the methods for this colour measurement system for human colour vision (Wright, 1969; Wyszecki & Stiles, 1967).

Using three primary wavelengths (for humans) of 436, 546, and 700 nm, and the Plankian relationship, the relative amounts of energy an object emits in each wavelength for any given temperature can be calculated, and the proportion of energy in those wavelengths plotted on a trichromatic triangle. The equal-energy white point (E), being the locus of equal energy for each wavelength, falls at the centre of the triangle (Fig. 3).

Figure 3 also shows the chromaticity co-ordinates for the Plankian locus (the co-ordinates of a heated (black body) object emitting visible light) in the three

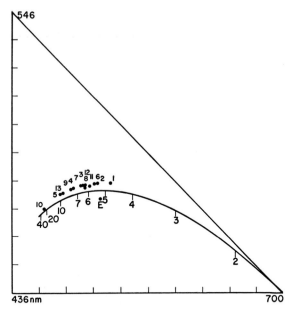

Figure 3. Chromaticities on an HVS colour triangle of the Plankian locus from 2000° to 40,000°K (large numbers along line) and daylight chromaticities from various sources (small numbers and points):

1, at 4800°K; 2, at 5500°K; 3, at 6500°K, 4, at 7500°K; 5, at 10,000°K (from Judd *et al.*, 1964); 6, at 5500°K; 7, at 6500°K (from Winch *et al.*, 1966); 8, at 6500°K (from Henderson & Hodgkiss, 1963); 9, at 7500°K (from Winch *et al.*, 1966); 10, at 34,000°K; 11, at 6020°K; 12, at 6440°K; 13, at 10,300°K (from Henderson & Hodgkiss, 1963). E is the equal energy white point.

primary wavelengths chosen to represent human colour perception. Also shown are some daylight chromaticities calculated from normalised spectral energy distributions. The closeness with which the region of daylight chromaticities parallels the Plankian locus shows that for human colour perception, daylight and Plankian radiation are almost identical, and that the equal energy white point (E) is sufficiently close to a commonly encountered natural light to make that white point a useful standard in colour measurement.

COLOUR VISION, SPECTRAL SENSITIVITY, AND DAYLIGHT IN THE IVS

Trichromatic colour perception by insects

Reviews on insect colour vision point out the similarity of the spectral sensitivity of insects (Mazokhin-Porshnyakov, 1969; Goldsmith & Bernard, 1974, and others). Most research has been on honeybees (*Apis mellifera* (L.)), and serves as the foundation of the theory I wish to develop.

Kuhn (1927) showed that the honeybee's visual spectrum is shifted entirely towards the shorter wavelengths of the daylight spectrum (*c.* 300–700 nm). Daumer (1956) developed a technique to exclude wavebands of light, or alter the brightness of those wavebands and measure the effects on the behaviour of trained honeybees. From that he devised a trichromatic colour perception model for honeybees. His results dovetail closely with electrophysiological studies

(Autrum, 1965, 1968). Mazokhin-Porshnyakov (1962) proposed trichromatic colour vision in bumblebees (*Bombus* spp.). Recent work by Högland *et al.* (1973a, b) and Struwe (1972a, b) suggest a trichromatic system of colour receptors in some Lepidoptera. Many workers have found three or more sensitivity maxima for the receptors of the compound eye of a wide variety of insects (see Goldsmith & Bernard, 1974).

For this study, I will use the three primary wavelengths of 360, 440, and 588 nm to define the chromaticity diagram and reference colorimetry in the insect visual spectrum (IVS). These wavelengths correspond to those considered by Daumer, Mazokhin-Porshnyakov and others as primary colours to insects (*vide* Goldsmith & Bernard, 1974). The use of similar reference primary wavelengths drawn from other sources would make no difference to the general considerations which follow.

Daylight in the IVS

I am not aware of research that has investigated the relationship between (i) insect colour perception, (ii) the daylight spectrum, and (iii) the Plankian locus. It is interesting to plot the Plankian radiator locus and representative daylight chromaticities on a trichromatic triangle befitting the IVS. The results given in Fig. 4 stem from calculations of the theoretical Plankian relationship and from examinations of daylight spectral energy distributions in the same manner as for the HVS but using IVS primary wavelengths.

From these calculations and observations, one can readily see that the Plankian locus is no longer an approximation to daylight as in the HVS; the daylight chromaticities take on a wide area of the triangle extending towards the apex.

Particularly important to developing this theory is the inappropriateness of the concept of an equal energy white point. That point is well removed from the

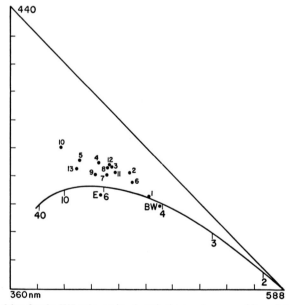

Figure 4. As Fig. 3 but for the IVS colour triangle. E is the equal energy white point. BW is bee white after Daumer (1956) (see text for explanation).

Plankian locus and lies even further from daylight chromaticities. Daylight *must* be regarded as the natural light to which diurnal insects are exposed.

It is noteworthy that much behavioural and electrophysiological research on insect vision has used or assumes equal energy stimulation. Daumer (1956) and Mazokhin-Porshnyakov (1962, 1969) show equal energy white points in their figures. This is clearly invalid for trichromatic colour vision in bees.

Spectral sensitivity of insects and daylight

The spectral sensitivities and colour vision of insects have been reviewed extensively (von Frisch, 1960; Goldsmith, 1961; Burkhardt, 1962, 1964; Mazokhin-Porshnyakov, 1969; Goldsmith & Bernard, 1974), but not apparently in direct apposition to daylight. Spectral sensitivity curves, whether obtained from data from mass electrical responses of the eye or from behavioural studies are similar—they show greatest sensitivity in the ultraviolet (UV) and a lesser peak of response in the green to yellow-green part of the spectrum.

Much attention is being paid to single cell spectral sensitivity. However, in this study, the behavioural approach is more appropriate to insect colorimetry as all the responses from the transducers of the eye to the central nervous system, and resultant behaviour are included. Mass electrical responses are also appropriate, as ganglionic integration of information is part of the data gathered. Simultaneously, this electrophysiological technique may remove confusion which could result from temporary colour preferences mediated by feeding, oviposition, or mating requirements.

Single cell recordings are invaluable in the understanding the mechanisms of colour vision, but do not seem to lend themselves to apposition with natural light spectra.

I want to demonstrate that the insect spectral sensitivity, particularly of bees, is adapted very closely to the quality of light in their environment. It is reasonable to suppose that where there is least light in the spectrum, (UV), there is greatest

Table 1. The relationship between the relative sensitivity of the bee's eye (S_λ) to the primary wavebands (M_λ) (after Daumer, 1956) and the amount of energy in those wavebands 4800°, 5500°, 6500°, 7500°K from Judd *et al.*, 1964). It is noteworthy that blue is the most stable for values of M_λ and S_λ. M_λ at various correlated colour temperatures of daylight, and that blue light is involved in the control of physiological adaptation (Menzel, 1971)

Waveband (nm)	Colour	Relative stimulation (Daumer, 1956)	S_λ normalised column 3	Per cent of energy, M_λ at daylight CCT: 4800°K	5500°K	6500°K	7500°K	Solar
300–390	Ultraviolet	5.6	0.1163	7.6	9.3	13.8	16.6	16.8
410–480	Blue	1.5	0.0317	28.7	29.2	33.2	33.4	30.0
500–650	Yellow	0.8	0.0167	63.7	61.9	53.0	50.1	54.2
				Values of S_λ . $M_\lambda \simeq 1$				
				0.88	1.08	1.60	1.93	1.91
				0.91	0.93	1.06	1.06	0.96
				1.06	1.05	0.90	0.85	0.92

sensitivity, and where there is the most light there is least sensitivity. The relationship can be expressed:

$$S_\lambda . M_\lambda = 1 \qquad \text{where S is sensitivity,}$$

M is spectral concentration,

λ is wavelength of primary colour.

Table 1 supports the validity of this relationship.

COLORIMETRY IN THE IVS

White

As pointed out above, equal energy white (appropriate to human colour vision) is inappropriate to insect colour vision in natural light. I want, therefore, to erect a rigorous ontological *definition of white* which is appropriate to colorimetry using any primary wavelengths.

Definition

First: *A perfectly white surface is one which is a perfect diffuser reflecting all the wavelengths impinging on it.* That definition takes no account of the sensitivity of the sensor 'seeing' white and further perfect diffusers and total reflection are theoretical. Also, the light source is ignored. To account for those difficulties the following definition is given: "*An object, the colour of which falls on the white locus of a chromaticity diagram, constructed for a given colour sensor, is a diffuser which reflects in equal proportions all wavelengths in (i) a spectrum of natural light within the range of sensitivity of that given colour sensor or (ii) the primary colours of that colour sensor.*" For this study, and most other colorimetric studies, a trichromatic sensor and chromaticity diagram are appropriate, and the natural light is assumed to be daylight. The definition is not unequivocal, as the spectrum of natural light, including daylight, is variable. Colour perception may also change as a result of adaptation of the eyes (Crawford, 1959; Autrum, 1968; Thomas & Autrum, 1965; Ripps & Weale, 1969; Menzel, 1971). Nonetheless, it should be noted that all primary wavelengths must be present in unnatural spectra in amounts above some threshhold for adaptation, to accommodate the balancing of the relatively weak wavelength(s) against the relatively strong wavelength(s). Presumably adaptation of colour perception occurs through comparison of reflected light to ambient lighting.

The above definition, based on *equiproportionate reflection*, avoids the equal-energy white standard, accommodates colour measurement in any spectrum of light approximating the natural light spectrum of the sensor, and allows for colour measurement independently of spectral adaptation of the sensor.

Colorimetry

With a definition of white which dictates that its position in the chromaticity diagram falls at the centre, colour analyses in both IVS and HVS fall into place. By simple analogy to elementary colorimetric methods (Wright, 1969; Wyszecki & Stiles, 1967) colours in both visual spectra can be measured. Table 2 shows how

Table 2. Trichromatic constructions of some colours to insects and man (see Figs 5 and 6)

Colours	1 A	1 B	1 C	2 A	2 B	2 C	3 A	3 B	3 C	4 A	4 B	4 C	5 A	5 B	5 C	6 A	6 B	6 C	7 A	7 B	7 C
Insect-blue blue	80	50	—	40	33	—	0	0	—	0	0	—	80	33	—	0	0	—	50	33	—
insect-green green	0	0	0	0	0	0	0	0	0	80	50	33	80	33	33	80	100	50	20	13	13
insect-red red	80	50	50	80	67	50	80	100	50	80	50	33	80	33	33	0	0	0	80	53	53
red	80	—	50	80	—	50	80	—	50	80	—	33	80	—	33	80	—	50	50	—	33

A is the % of available light reflected in that wave-band
B is the proportion of that wave-band making the colour for insects
C is the proportion of that wave-band making the colour for man
1 is insect-purple
2 is insect red-purple
3 is insect-red
4 is insect-yellow

5 is insect-white
6 is insect-green
7 is insect-mauve
1, 2, 3 are all yellow
4, 5 are both white
6 is purple
7 is a greenish yellow

the proportionate reflection in the three primary wavebands, for man and insects, is used to determine the colour of seven representative surfaces. Figures 5 and 6 illustrate how these colours are placed on human and insect trichromaticity diagrams. Plates 2 to 5 illustrate the same for floral colours.

Colour naming

Daumer (1956, 1958) introduced a number of special insect-colours, namely "bee-violet", "bee-purple", and "bee-white", in conjunction with names which we attach to colours. This creates a confusing inconsistency, which compounds the difficulty in appreciating the range of colours as they might appear to insects, *I propose that colours be named for insects analogously as for man*, that is colours at the same analogous positions on the insect trichromatic diagram receive the

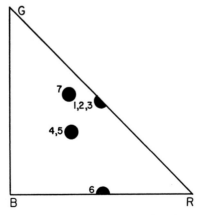

Figure 5. Trichromatic plots for the HVS colour triangle of the colours described in Table 2 and by their spectral reflectance curves in Fig. 7A and B. Points 1, 2 and 3 are yellow, 4 and 5 are white, 6 is purple, 7 is greenish yellow.

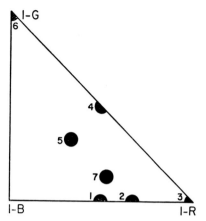

Figure 6. Trichromatic plots for the IVS colour triangle of the colours described in Table 2 and by their spectral reflectance curves in Fig. 7A and B. Point 1 is insect-purple, 2 is insect-reddish-purple, 3 is insect-red, 4 is insect-yellow, 5 is insect-white, 6 is insect-green, 7 is insect-mauve.

same name (compare Figs 5 and 6), but merely prefixed by the word 'bee' or 'insect' (Kevan, 1972a). Illustrations of this naming system are given in Table 2, thus, what we call yellow is "insect-red"; blue is "insect-green"; ultraviolet is "insect-blue"; and Daumer's other special bee-colour names remain valid.

METHODS

Parameters

Most research on the "colours" of objects in the IVS is, unfortunately, restricted to only the presence or absence of UV. As demonstrated by Daumer (1956, 1958) and Mazokhin-Porshnvakov (1966, 1969), insects, especially bees, can see a large array of colours. To measure colours in the IVS, it is not enough to record the presence or absence of UV (Table 2). Recording the UV reflectance lacks meaning as does recording *in vacuo* green reflectance. An object reflecting green could be bee-white through to bee-dark-grey, bee-green, bee-yellow, bee-bluegreen; and any degree of saturation (see below) thereof. The sensitivity of insects' eyes to lights of various wavelengths, and insects' apparent appreciation of a wide array of colours, requires that proper colorimetry be done if a complete understanding of the significance of colours to insects is to be approached (Kevan, 1972a; Kevan *et al.*, 1973).

To accomplish this within the principles of colorimetry, *a spectral reflectance curve* must be generated to show the percentage of the available light reflected (*luminance factor*) in each wavelength across the spectrum of interest, here the IVS. From that, the colour of the object can be represented on a trichromaticity diagram. Spectral reflectance curves for colours in Table 2 are shown in Fig. 7.

The spectral reflectance curve also shows the *hue*, which, broadly speaking, is determined by the spectral position of the more strongly represented wavelengths; and the *saturation*, governed by the amount to which the stronger wavelengths predominate (Fig. 7). The luminance factor and saturation considered together give information on the *paleness* and *dullness* of colour: as saturation decreases, the stronger wavelengths become less dominant; and as the luminance factor increases, the colour becomes paler; as saturation decreases with luminance factor, the colour becomes duller (see Fig. 7). The luminance factor and hue taken together give information on the *brightness* and *darkness* of colour: the greater the luminance factor, the brighter the colour. (I have avoided the term *lightness* as it confuses paleness with brightness.) Paleness, dullness, brightness,

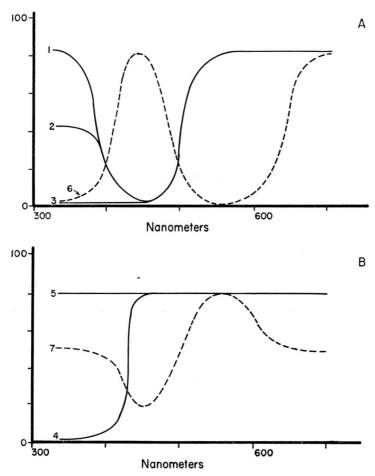

Figure 7. A. Spectral reflectance curves for colours 1, 2, 3, and 6 of Table 2 and Figs 5 and 6. B. Spectral reflectance curves for colours 4, 5, and 7 of Table 2 and Figs 5 and 6.

Curve or colour no.	Insect colour	Human colour
1	purple	yellow
2	reddish purple	yellow
3	red	yellow
4	yellow	white
5	white	white
6	green	purple
7	mauve	greenish yellow

and darkness cannot be represented on the trichromaticity diagram, as dark grey through to white fall on the white point. To accommodate co-ordination of all measurements, a colour-solid representation must be invoked (Wyszecki & Stiles, 1967), and is beyond the scope of this paper.

Instrumentation

Spectral reflectance curves can be obtained in several ways. Richtmyer (1923) and Lothmar (1933) used quartz spectrographs. Spectrophotometers are not well suited to field work, but are needed to determine the reflectance (luminance) of standards to be used in field work (e.g. grey-scales, Kevan et al., 1973) or of more or less homogeneously coloured surfaces (e.g. experimental objects to test colour preference).

Lutz (1924, 1933) used a pin-hole camera to avoid the problem of the opacity of glass to UV. A camera with a quartz lens and wide band monochromatic filters is simple to operate. A fast panchromatic black and white film (e.g. Kodak Tri-X; ASA 400) serves well. Daumer (1958), Kevan (1970, 1972a), and Mulligan & Kevan (1973) quantified the reflectance from flowers across the IVS using different photographic techniques. Mulligan & Kevan (1973) used a specially constructed and calibrated grey-scale (Kevan et al., 1973) included in photographs taken through a series of wide band monochromatic filters across the IVS and HVS (see Plates 2 to 5). One problem in filter photography in the field is correct exposure (Kevan, 1972a). Focus is another problem in photography at shorter wavelengths but can be overcome with special focusing filters available with some quartz lenses (e.g. Pentax) or chromatically adjusted lenses (e.g. Zeiss). Silbergleid (1976) recently reviewed the ways by which UV reflectances can be examined photographically. Some techniques are applicable to IVS and HVS colorimetry, particularly incorporation in the camera of a transmission grey scale, made of neutral density filter strips (Daumer, 1958).

Recently television cameras, e.g. video recorders, have been used with UV passing filters and quartz lens to examine UV reflections from flowers (Eisner et al., 1969; Gulberg & Atsatt, 1975; Jones & Buchmann, 1975). This versatile system can also be used with a calibrated grey-scale and wide band monochromatic filters. Kevan & Laverty (unpubl.) are using this, and other, techniques on the Colorado alpine flora and fauna.

FLORAL COLOURS IN THE IVS

Weeds of eastern Canada

Weed species are notorious for producing abundant seed by self-pollination (autogamy) or by agamospermy (Baker, 1974). Allogamous weeds are mainly self-incompatible perennials (Mulligan & Findlay, 1970). Further, Mulligan (1972) found that self-incompatible weeds are frequently visited by insects, whereas self-compatible weeds are relatively ignored. To determine whether there are special characteristics of weed flowers which attract anthophilous insects, Mulligan & Kevan (1973) studied floral attractants and included colorimetric analyses across the IVS and HVS.

I will review the colorimetric analyses here and expand the conclusions. Figure 8 shows the trichromatic plots for the colours of weed flowers in the IVS, and refers to Table 3 of Mulligan & Kevan (1973). Figure 9 shows the colours as plotted on a triangle for the HVS.

Table 3. Ultraviolet reflectance and frequency of visitors to flowers of Canadian weeds

Visitor frequency	No. of weed spp.	Number and per cent of flowers with:			
		High UV (>40%)	Medium UV (20–40%)	Low UV (<15%)	No UV (<5%)
1 (high)	14	4, 29%	6, 43%	1, 7%	3, 21%
2	24	4, 17%	4, 17%	6, 25%	10, 42%
3	9	1, 11%	2, 22%	2, 22%	4, 44%
4 (none)	6	2, 33%	0, 0%	3, 50%	1, 17%

Several points arise from these data. First (Table 3), there is only a slightly greater frequency of visitors to UV reflecting flowers; 33% of the unvisited flowers show high UV reflections, 29% of the well visited flowers show high UV, 17% and 21% of the flowers respectively show no UV reflection. The same general conclusion can be drawn if one similarly considers blue or yellow reflectance (see Mulligan & Kevan, 1973: fig. 24). Large flower size, and odour are more important.

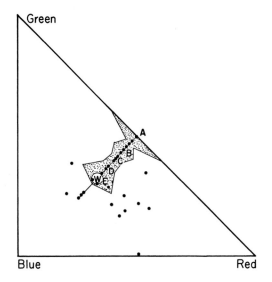

Figure 8. Trichromatic plots of the flower colours of Canadian weeds (from data of Mulligan & Kevan, 1973) in the HVS.

Letters A, B, C, D, and E are points shared by many species. The width of the shaded area is proportional to the number of observations at that colour point: A, 37 observations of yellow flowers or floral parts; B, 12 observations; C, 5 observations; D, 6 observations; E, about 22 observations. Other spots are single observations. W is the equiproportionate reflectance white point.

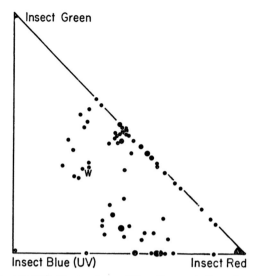

Figure 9. Trichromatic plots of the flower colours of Canadian weeds (from data of Mulligan & Kevan, 1973) in the IVS. Diameter of spots is proportional to the number of observations, mostly single. Largest diameter area at insect-red represents 37 observations. W is the equiproportionate reflectance white point.

From trichromatic plots in the IVS and HVS (Figs 8 and 9), one can appreciate the greatly expanded gamut of floral colours to insects. Floral colours are also more distinct to insects. In the HVS are many flowers or floral parts (50 observations) which fall in the yellow to pale yellow colour range, and a further 25 more observations around the white point. Distinct colours are shown in the purple region (7 observations) and to a lesser extent in the blue (5 observations). In the IVS plot, floral colours are not tightly clustered. Those reflecting insect white (5 observations) are well separated from those reflecting insect-yellow (19 observations), which are well separated from the insect-yellow-orange area (12 observations). The insect-purple area is spread out (23 observations) as are the insect-green (9 observations) and insect-orange areas (5 observations). The insect-red area is congested by observations (37) (HVS yellow without UV reflectance), but of these only 7 are of whole flowers, the remaining 30 being of insect-red floral parts in combination with other coloured floral parts. When all this is considered, one sees that almost every weed flower is a distinctive colour.

Some visitation/visitor and colour correlations appear. Bees (Apoidea) are more prevalent than other anthophiles on blue or purple flowers (insect-blue to insect-greens), such as *Vicia cracca*, *Malva neglecta*, *Cichorium intybus*, *Arctium minus* and others. Often blue coloured flowers are adapted to bee pollination, e.g. many Boraginaceae, Campanulaceae, Gentianaceae, Scrophulariaceae, and Polemoniaceae.

In general, blue reflecting flowers are not bright—rarely over 50% reflectance in any waveband. That blue flowers are especially attractive to bees is an old idea: Lubbock (1881) concluded that blue was bees' favourite colour. The idea would be better worded as 'blue reflecting flowers (including purple but excepting white) are more frequented by bees than by other insect groups'.

Blue reflecting flowers stand out from their backgrounds more from being brightly coloured differently (see below) than from being brightly visible. This contrasts with white (insect-yellow) coloured flowers which reflect brightly across the IVS and HVS. Generally, reflectance is above 50% and about 90% in *Lychnis alba* and *Stellaria graminea*. Parasitic Hymenoptera visit white flowers with easily accessible nectar, e.g. *Daucus carota*, *Stellaria* spp. (Kevan, 1973a).

Flowers of the Canadian high arctic

It was long thought that arctic flora consisted of autogamous inbreeders and polyploid agamosperms. However, recent data has demonstrated the importance of entomophily in high arctic plant reproduction (Kevan, 1972b; Shamurin & Tikhmenev, 1974). I (Kevan, 1972a) correlated the floral attractants for insects in high arctic flowers with insect visits, floral rewards, and sexual reproductive strategies of the plants (Kevan, 1970). In many respects the data can be compared directly with that of Mulligan & Kevan (1973).

Table 4. Ultraviolet reflectance and frequency of insect visitors to flowers in the Canadian high arctic

Visitor frequency	No. of species	Number and per cent of flowers with:			
		High UV (40%)	Medium UV (20–40%)	Low UV (15%)	No UV (5%)
1 (high)	7	1, 14%	0, 0%	2, 28%	4, 57%
2	6	0, 0%	1, 17%	1, 17%	4, 66%
3	9	1, 11%	0, 0%	3, 33%	5, 56%
4 (none)	8	0, 0%	1, 13%	0, 0%	7, 88%
				(*Braya* spp. and *Draba* spp. included as 1 each)	

Kevan (1972a) presents the colours of flowers of the high arctic. UV reflection is worth examining in the same way as for weeds (Table 4). UV reflections do not contribute greatly to the attractiveness of flowers. The trichromatic plots, Figs 10 and 11, show again the expanded gamut of distinct insect-colours. Generalisations on floral colours, reflectance, and visitation are difficult, despite the reduced diversity of flora and floral colours. The colours of anthia with high attractiveness range from the dullest (e.g. pistillate aments of *Salix arctica*) to the brightest insect-yellow (*Stellaria longipes*), from dull insect-yellow/green (*Saxifraga oppositifolia*) to bright insect-red (*Potentilla nivea*), insect-orange and red (*Dryas integrifolia*) and two-tone insect-purple and red (*Arnica alpina*). The same range of coloration is found in relatively unattractive flowers, including the intricately and brightly patterned *Saxifraga hirculus* (see Plate 1, Frontispiece).

Among attractive flowers, those reflecting blue received proportionately greater attentions from bumblebees than from other insects, and the bright insect-yellow *Stellaria longipes* the attentions of parasitic Hymenoptera. The yellow flowers, whether reflecting UV or not, are visited by almost all anthophiles.

Kevan (1970, 1973b) suggests that, as with weeds, anthia size and floral scent can be more closely related to floral attractiveness. Recent work on the north slope of Alaska corroborates these general conclusions (Kevan, 1976a).

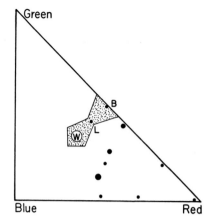

Figure 10. Trichromatic plots of the flower colours of the Canadian high arctic flora (from data of Kevan, 1970, 1972a) in the HVS.

Letters B, L, and W are points shared by many species. The width of the shaded area is proportional to the number of observations at that colour point: B, 22 observations of yellow flowers or floral parts; L, 6 observations of pale yellow flowers or floral parts; W, 20 observations of white flowers or floral parts. Diameter of spots is proportional to the number of observations. W is the equiproportionate reflectance white point.

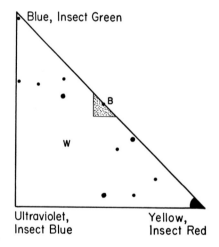

Figure 11. Trichromatic plots of the floral colours of the Canadian high arctic flora (from data of Kevan, 1970, 1972a) in the IVS.

Diameter of spots is proportional to the number of observations, smallest for single. Largest diameter area at insect-red represents 14 observations. Shaded area B represents 13 observations. W is the equiproportionate reflectance white point.

Colours and other floral attributes

Despite the well founded observations that UV is highly visible to insects, it appears no more important in attraction than other wavebands of light reflected from flowers. Full colorimetric analyses of flowers do not allow the prediction of attractiveness of a flower to insects generally, or specifically. This does not imply that floral coloration is unimportant in attracting insects, quite the contrary.

Colour is an important floral attribute. It functions as a long distance signal especially to actively flying diurnal anthophiles, e.g. Diptera and Hymenoptera. The array of floral colours visible to insects allows them to distinguish species of flowers at a distance. This is important when one considers that the resolving power of the insect eye is inferior to our own (Hocking, 1964; von Frisch, 1965; Goldsmith & Bernard, 1974). Thus, as we recognize flowers of the same colour from a distance by shape; insects, particularly bees, can recognise the indefinitely resolved shape at distance by its distinctive colour.

Colour is part of the array of visual attractants flowers possess. Generalisations on associations of anthophiles and floral colours are well known (Faegri & van der Pijl, 1971; Proctor & Yeo, 1973). I have remarked on blue reflecting flowers and bees, but bee-flies (Diptera: Bombyliidae) are associated with blue flowers (Knoll, 1921), as are some more advanced butterflies (Ilse & Vaidya, 1956) visiting flowers of *Dianthus, Phlox, Eupatorium, Cirsium, Lychnis dioica*, and others (Proctor & Yeo, 1973). White flowers with open nectar are visited by Parasitica (Hymenoptera) (Kevan, 1973a). Better known are white nocturnally blooming flowers with their appropriate heavy characteristic odours in psychophily and chiropterophily.

Yellow flowers are often highly reflective and attract every conceivable diurnal anthophile. Unspecialised anthophiles show preference for yellow (Ilse, 1949; Ilse & Vaidya, 1956; Kugler, 1951) as do less advanced butterflies, e.g. Nymphalidae (Ilse, 1928; Tinbergen *et al.*, 1942). Valentine (1975) concludes that yellow floral colour is adaptively neutral. The red hummingbird flower pollination syndrome is well known (Grant, 1966; Grant & Grant, 1968; Raven, 1972). In the Mediterranean region, the red flowered *Lychnis chalcedonica* is visited by butterflies, as are many red flowered tropical and subtropical plants (Proctor & Yeo, 1973). It is noteworthy that some advanced Lepidoptera have been found to have red sensitive vision (Goldsmith & Bernard, 1974). As yet the full colour vision capability of these insects is unknown and we have no full colorimeteric analyses of the flowers they visit.

Mimetic coloration of dung and carrion scented blooms is well known (but not colorimetrically investigated). Kullenberg (1956, 1961) discusses the role of colour patterning (including UV) of *Ophrys* flowers in pollination through copulation attack by the Hymenoptera they mimic in colour and scent. Both flowers and insects are similarly coloured across the IVS, except that the mirror patches of the flowers are more reflective in UV than the wings of *Gorytes* (Hymenoptera: Sphecidae): The labellum of the flower is, then, a "super-normal visual stimulation" (1961: 294).

Size is an important characteristic (Darwin, 1877). Kugler (1943) showed that the distance at which a blossom visually attracts bumblebees is directly proportional to its diameter. Horovitz & Harding (1972) point out the danger in generalisation of this point. In *Lupinus nanus* some highly self-pollinating forms have large flowers, whereas forms with greater amounts of outcrossing have smaller flowers. On the other hand, Lewis (1965) provides an example of diminished flower size with the evolution of selfing. Frequently, agamospermous and autogamous flowers of weeds and arctic plants have large, colourful flowers or anthia that are well visited by anthophiles.

Floral guides

Floral guides are another form of close-in attractant. They may be olfactory, structural, or visual.

Sprengel (1793) first described the function of nectar guides as markings which help anthophiles orient on the flower to reach the reward sought—nectar, or pollen, or both. Predictably, nectar guides are more common on flowers with hidden rewards, as noted by Müller (1876) in *Saxifraga* spp., and especially in zygomorphic flowers by Kugler (1930, 1963). The guides may be visible to humans, e.g. the lines in flowers of *Viola* spp., the orange lip of *Linaria vulgaris*, the yellow centre of *Myosotis*, the pattern of spots on *Digitalis purpurea*, the patterns in *Iris, Delphinium tricorne* (Macior, 1975; Utech & Kawano, 1975), many orchids, many legumes, etc. These patterns may have a UV component. Thein (1971) shows guides on orchids which disappear in UV light, as does the pattern on *Linaria*. Some guides are invisible to the human eye; reflecting UV light, they are visible to insects (e.g. on *Lupinus nanus* (Horovitz & Harding, 1972), *Saxifraga hirculus* (Kevan, 1972a)).

The term nectar guide is also applied to the concentric colour effects brought about by UV (and other) reflections. These reflections are not precise guides, but as Daumer (1958) observed, the pattern is important in bees' orientation towards the nectar or pollen; i.e. bees are acutely aware of positional relationships of the UV reflecting area and the UV absorbing area (texture could also be involved here, see Brehm & Krell, 1975). Watt, Hoch & Mills (1974) suggest that *Colias* butterflies also orientate to target patterns of some Rocky Mountain Compositae. The two-colour target may have a bull's eye which is relatively large, e.g. *Arnica alpina* (Kevan, 1972a), *Ranunculus* spp., *Taraxacum* spp. (Kevan, 1972a; Mulligan & Kevan, 1973; Daumer, 1958; Mazokhin-Porshnyakov, 1959; Utech & Kawano, 1975), or highly restricted, e.g. some Cruciferae (Horovitz & Cohen, 1972), *Oenothera biennis* (Daumer, 1958; Mulligan & Kevan, 1973; Utech & Kwano, 1975), and *Nymphoides indica* (Ornduff & Mosquin, 1970). Two-colour target patterns are not necessarily invisible to humans, as is well demonstrated in *Erigeron* spp., *Chrysanthemum* spp., *Aster* spp., *Myosotis* spp., *Dryas integrifolia*, and so on. In target patterns, the shortest waveband is reflected peripherally and insect-red (mainly) or insect-yellow, the colours to which insects have the least amount of hue discrimination, make up the centre.

Nectar guides neither indicate attractiveness to anthophiles, nor give clues as to sexual reproductive strategies. *Taraxacum* sect. *Vulgaria* are agamospermous and well visited, as is the arctic *T. arctogenum; Saxifraga hirculus* is probably agamospermous and is not well visited. *Verbascum thapsus* (Mulligan, 1972) and *Eritrichium aretiodes* (Petersen, 1968) are autogamous, but visited by insects. Various reproductive strategies are seen in *Ranunculus, Potentilla, Viola,* and *Taraxacum,* and in the complex species *Lupinus nanus* (Horovitz & Cohen, 1972) and *Nymphoides indica* (Ornduff & Mosquin, 1970), along with various nectar guide patterns.

An interesting "nectar guide" phenomenon is described by Knoll (1922) in *Muscari comosum,* in which the upper infertile flowers of the raceme are violet and attract insects, whereas the lower, fertile flowers are brownish yellow, provide nectar, and are visited.

Kugler (1963) examined the occurrence of visible (in the IVS) and UV patterns in different forms of flowers. Butterfly flowers showed the highest incidence of guides (83%), both UV (63%) and visible (66%). Next were lipped flowers, 76% with guides, showing a greater prevalence of visible (68%) patterns compared with UV (42%). Salverform and capitulate forms also showed high incidence of guides (74% and 67% respectively), but with UV being more prevalent (56% *vs.* 30%; and 61% *vs.* 36% respectively for UV *vs.* visible). Funnel, disc, and bell forms had about even numbers with and without guides, showing a slight tendency in the first two forms towards UV patterns.

Patterning is clearly a feature of advanced floral forms, but within them, neither UV nor visible patterns give further firm indications of pollination specialisation.

The point is that any floral attribute, including nectar guides, has a two-fold function: (1) to help the anthophile obtain reward, and (2) to help the plant reproduce sexually. The importance of nectar guides is well demonstrated for the first function (Knoll, 1921; Kugler, 1930, 1938; Clements & Long, 1923; Daumer, 1956, 1958; Manning, 1956a, b; Free, 1970; Jones & Buchmann, 1975), but data for the second function seem scanty.

Nectar colours

Recently Thorp *et al.* (1975) have suggested that nectar itself may absorb, reflect, or fluoresce daylight, acting as another visual attractant. Specular (mirror-like) reflection would indeed be highly visible to visiting anthophiles, the nectar shining as glistening droplets in open flowers. Kevan (1976b) has suggested that coloured light emissions from nectar would be too weak, in the complex colour and shade environment of a flower, to be perceived by anthophiles, yet the correlation of fluorescent nectar being more frequent in open flowers than in flowers with hidden nectar bears consideration. Further research is underway by Thorp and his co-workers (see Thorp *et al.*, 1976) to see if another item must be added to the list of floral attractants.

Flower and floral phenology

In some flowers, colours may change through the life cycle. Daumer (1958) noted that buds of *Chelidonium majus*, unlike open flowers, did not reflect UV. Eisner *et al.* (1973a) have discussed the UV patterns of buds and blossoms. They described buds of *Jasminium primulinum* and of *Hypericum* spp. as being UV absorbing, and as the buds open, the newly exposed petal parts are UV reflecting. The resultant pattern on the back of the flowers (wedges of insect-purple and of insect-red) does not match the coloration of a typical target pattern on the front.

Other flowers change colour as they age. The familiar Ragwort (*Senecio jacobaea*) capitulum turns brown at the central florets, while the ligulate florets remain yellow. See also Plate 4. Kugler (1950) found the brown centred flowers were ignored by hover flies (*Eristalis tenax*). In the Leguminosae, ageing of the flowers is often accompanied by fading and browning. In *Oxytropis splendens* in the Colorado Rockies, Laverty (unpubl.) has found that the banner petals fade, brown and wither first, and correlates this with reduced visitation of bumblebees. The horse-chestnut (*Aesculus hippocastanum*) has flowers which change noticeably, having first yellow spots which become red with age and do not then attract pollinators (Kugler, 1936).

The flowers of *Mertensia* spp. change colour from bud to maturity, starting out pale pink and becoming increasingly blue, resulting from the sensitivity of the anthocyanin pigments to changes in pH (Weevers, 1952). The opposite trend, from blue–violet to purple–red is found in *Fuchsia hybrida* as pH declines (Yajaki, 1976). Laverty (unpubl.) has noted an increase in the frequency of bumblebee visits with increase in the blueness of the flowers which signals greater amounts of nectar and pollen being present. Many other Boraginaceae have similar colour changes.

Many flowers undergo changes following pollination. Jones & Buchmann (1975) note that in *Caesalpinia eriostachys*, the banner folds down after anthesis, changing the floral symmetry and colour pattern. In *Parkinsonia aculeata* the colour with the banner becomes increasingly orange. Orchids are notorious for withering immediately after pollination.

One South African *Gladiolus grandis* subspecies remains brownish during the day, but is blue at night (Ingram, 1967). Henslow (1893) records a *Phlox* which starts at dawn as pale blue but becomes pink by mid-morning.

Horovitz & Harding (1972) suggest a role for nectar guides and coloration in seasonal events. Comparing two genetic stocks of *Lupinus*, they suggest that the relative lack of blue reflectance and nectar guide spots on one accounts for the lower outcrossing rates early in the season, when attractive devices are critical. Later, when pollinators were more abundant, outcrossing was higher. Perhaps, the old fashioned idea of floral colour seasons (see Kerner & Oliver, 1904, II: 197–198) should be resurrected and subjected to critical examination through IVS colorimetry and seasonal abundance of anthophiles.

Crypsis in anthophiles, and protective coloration of flowers

Predators and prey species may be cryptically coloured. This appears true for predatory anthophiles (e.g.: crab spiders (Araneida: Thomisidae)) which are camouflaged in their flowers. Eisner *et al.* (1969) show an UV reflecting crab spider (possibly *Thomisus formosipes*) on the UV absorbing disc of *Viguiera dentata*, where it would be conspicuous to potential prey or predators. Kevan (1972a) recorded the arctic crab spider, *Xysticus deichmanii* (UV absorbing) in flowers. On *Arnica alpina* they were never found at the distal ends of the ligulate florets where they would contrast. Specimens of *X. deichmannii* taken from purple flowers had a purple tinge rather than the yellow tinge of those from like coloured flowers. Wickler (1968) records the crab spider, *Misumena vatia*, as able to assume different cryptic colours according to its background.

Some potential prey anthophiles also show cryptic coloration. Kevan & Kevan (1970), suggest that light coloured Collembola visit flowers in which they are inconspicuous. Anthophilous thrips are frequently light coloured, as are many Heteroptera. However, this aspect of coloration in anthecology remains almost unexplored.

Hinton (1973a, b) tantalisingly suggests that bright floral colours are warning colours. He argues that flowers should attract insect pollinators, but not herbivores which would eat the entire flower. Hence, bright floral colours and plant toxins in concert would have been selectively favoured in the Cretaceous when iguanodon and ceratopsian reptiles, presumably (as today's reptiles) with good colour vision, were the major herbivores.

Background colours

Flowers do not bloom against black velvet, the backdrop for much floral photography, but against a background of natural vegetation, sky or soils. Frohlich's (1976) considerations on the backdrop against which flowers are seen by insects have short-comings, but his points bear recognition. He considers only UV. Green vegetation reflects fairly evenly, and rather dully, across the IVS making the vegetation greyish, with tints of yellowish and bluish (Daumer, 1958; Kevan, 1972a) against which most flowers will contrast vividly (absorption of red light imparts the green appearance in the HVS). The contrast is less for purple and blue flowers, as noted. Soils in general are not highly reflective in UV (Condit, 1971). The sky is a bright backdrop in all wavebands. Thus, in the IVS (or AVS) most flowers are brighter than soil or vegetation and darker than the sky. Some flowers are more brightly coloured than others, and their contrast against vegetation has been discussed. Full colorimetric contrast can not be understood in a single waveband of the IVS. Two spectral reflectance curves are needed, one of the object and one of the backdrop.

Colours as isolating mechanisms

Understanding insect colours of flowers allows us to examine floral colours as isolating mechanisms. Varietal and genotypic differences in plants affect their attractiveness to pollinators, although not necessarily by colour (Free & Spencer-Booth, 1965; Free, 1970; Forster & Levin, 1967; Dennis & Haas, 1967; Moffett et al., 1975; Faulkner, 1976).

In the HVS, Levin (1969, 1972) and Levin & Schaal (1970) have shown that floral colours in *Phlox* spp. act as an isolating mechanism in reducing interspecific and intervarietal pollen flow, and inhibiting intra-specific hybridisation. Cruden's (1972) study of *Nemophila menziesii*, in which he studied differences in UV reflection of geographic races in relation to pollination by bees (mainly *Andrena*), shows that geographic races and plant speciation can be associated with pollinators discriminating on the basis of floral colours. However, Carter's (1974) study suggests the breakdown of isolating mechanisms in *Cercidium* spp. (Leguminosae) despite distinct coloration in the IVS. Ornduff & Mosquin (1970) examined reflection from flowers of *Nymphoides indica* complex from various parts of the world, concluding (admittedly without field evidence) that the evolution of contrasting floral patterns represents increasing specialization in pollination. Spotless mutants of *Monarda punctata* when growing amongst normal flowers are apparently ignored by their hymenopteran pollinators (Scora, 1964). Other examples of the effects of colour varieties in the HVS on foraging by anthophiles are discussed by Clements & Long (1923), Clement (1965), Free (1970), Kauffield & Sorensen (1971), Leleji (1973), Mogford (1974), Goplen & Brandt (1975), Faulkner (1976), and Kay (1976).

Kay's study shows strong preference of *Pieris* spp. and *Eristalis* spp. for yellow morphs of *Raphanus raphinastrum* which have the typical insect purple and insect-red target pattern. The white morph's pattern is not given; however, the petals reflect ultraviolet strongly, and are insect-white.

An important phenomenon in this framework is flower constancy or fidelity (Grant, 1950; Free, 1966). Insects recognise flowers by colour, shape, smell, and form. Bees, foraging at a species with more than one colour of flowers will forage with little discrimination (Darwin, 1876, 1877; Mather, 1947); but will ignore any peculiar individual (Scora, 1964; Free, 1966). Floral constancy is important to Cruden's (1972) study. Sympatric or parapatric speciation of an obligate zoophilous plant implies some isolating mechanism, unless the plant becomes more or less instantly reproductively isolated through polyploidy. Isolation through pollinator behaviour is discussed by Heslop-Harrison (1958) and Grant (1949).

Floral colours in taxonomy

Floral colours are often used as taxonomic characters, but IVS colorimetry has not been used. Eisner *et al.* (1973a) show that UV patterns are preserved on herbarium specimens. Horovitz & Cohen (1972) suggest that differences in UV reflections of closely related species (Cruciferae) may be useful taxonomically. Indeed, Carter (1974) has used qualitative IVS floral colour characters in her study of *Cercidium*.

Knowledge of the biochemistry of floral pigments and their distribution in flowers is increasing. Thompson *et al.* (1972) discuss the role of flavonols as UV absorbing pigments in *Rudbeckia hirta*. Anthocyanins have been used in chemo-taxonomic studies of plants (Arditti, 1969; Harborne, 1963, 1967). As more is learnt about the roles of floral pigments in nature, two fields which are presently separate, anthecology and chemotaxonomy, may share common ground. A parallel situation is the work of Kullenberg (1973), Kullenberg & Bergström (1976), Preisner (1973), and Williams & Dodson (1972) on orchid scents and bee phero-mones.

Biogeographical implications

Biogeographic studies of floral colours have not yielded particularly interesting results. Woodson (1964), in studying *Asclepias tuberosus*, was unable to suggest any selective pressure to account for biogeographical variation in flower colours in the HVS. Weevers (1952) discredits any theory of biogeography of floral colours, but did not consider arctic floras. Kevan (1972a) points out the relative abundance of yellow and white flowers in the arctic. Utech & Kawano (1975) corroborate this with data from Japan. They also introduce the interesting fact of the increased atmospheric ozone northwards, and the ozone's greater seasonal variation in the north. They link this to UV intensity, and suggest that its effect as a selective pressure of northern floras for UV absorption. The idea is interesting, but not likely. Selective pressures for floral colours presumably result from pollination mechanisms, and insects can readily adapt their colour vision to changes in daylight throughout the day and the year. Also, the relative amounts of ultraviolet in arctic daylight does not appear to differ greatly from elsewhere (Hisdal, 1967; Caldwell, 1972; Henderson, 1970). In montane regions UV is considerably stronger, and highly reflective floral parts may be advantageous for protective reasons (Caldwell, 1971). Unfortunately, few data exist on UV reflections or total colorimetry for entire regional floras to examine the idea further.

ACKNOWLEDGEMENTS

I thank G. A. Mulligan for use of the data on eastern Canadian weeds, and T. Laverty for unpublished results of his research in the Colorado Rockies. I would like to thank by name all those with whom I have corresponded on the subject of colour measurement, however, the list would be long. Instead, I would like to express my thanks to those people as a group.

I am grateful to A. J. Richards, the Linnean Society of London, and the Botanical Society of the British Isles for the opportunity of presenting my data and views at this conference. Funds for travel to the U.K. were also provided by my National Science Foundation grant No. DEB 76–20125 and by University of Colorado, Colorado Springs.

R. D. Leggett prepared the line illustrations, D. A. Davidson the black and white plates, and Gael Bennett the coloured plate, and I should also like to thank them.

REFERENCES

ARDITTI, J., 1969. Floral anthocyanins in species and hybrids of *Broughtonia*, *Brassavola*, and *Cattleyopsis* (Orchidaceae). *American Journal of Botany, 56:* 59–68.

AUTRUM, H., 1965. The physiological basis of colour vision in honeybees. In A. V. S. de Reuk & L. Knight (Eds), *Colour Vision Physiology and Experimental Psychology:* 286–300. CIBA Foundation Symposium. London: Churchill.

AUTRUM, H., 1968. Colour vision in man and animals. *Naturwissenschaften, 55:* 10–18.

BAKER, H. G., 1974. The evolution of weeds. *Annual Review of Ecology and Systematics, 5:* 1–24.

BOLWIG, N., 1954. The role of scent as a nectar guide for honey bees on flowers and an observation on the effect of colour on recruits. *British Journal of Animal Behaviour, 2:* 81–83.

BREHM, B. G. & KRELL, D., 1975. Flavonoid localization in epidermal papillae of flower petals: a specialized adaptation for ultraviolet absorption. *Science, 190:* 1221–1223.

BURKHARDT, D., 1962. Spectral sensitivity and other response characteristics of single visual cells in the arthropod eye. In J. W. L. Beament (Ed.), *Biological Receptor Mechanisms. Symposium of the Society for Experimental Biology, 16:* 86–109.

BURKHARDT, D., 1964. Colour discrimination in insects. *Advances in Insect Physiology, 2:* 131–173. London: Academic Press.

BUTLER, C. G., 1951. The importance of perfume in the discovery of food by the worker honeybee (*Apis mellifera* (L.)). *Proceedings of the Royal Society (Ser. B., Biological Sciences), 138:* 403–413.

CALDWELL, M., 1971. Solar UV radiation and the growth and development of higher plants. *Photophysiology, 6:* 131–177.

CALDWELL, M., 1972. Biologically effective solar ultraviolet irradiation in the arctic. *Arctic and Alpine Research 4:* 39–43.

CARTER, A. M., 1974. Evidence for the hybrid origin of *Cercidium sonorae* (Leguminosae; Caesalpinicidea) of Northwestern Mexico. *Madroño, 22:* 266–272.

CLEMENT, M. W., 1965. Flower color, a factor in attractiveness on alfalfa clones for honeybees. *Crop Science, 5:* 267–268.

CLEMENTS, F. C. & LONG, F. L., 1923. Experimental pollination. An outline of the ecology of flowers and insects. *Carnegie Institution Publication* (Washington), No. 336: vii+274 pp.

CONDIT, H. R., 1970. The spectral reflectance of American soils. *Photogrammetric Engineering, 36:* 955–966.

CRAWFORD, B. H., 1959. Measurement of color rendering tolerances. *Journal of the Optical Society of America, 49:* 1147–1156.

CRUDEN, R. W., 1972. Pollination biology of *Nemophila menziesii* (Hydrophyllaceae) with comments on the evolution of oligolectic bees. *Evolution, 26:* 373–389.

DARWIN, C., 1876. *Cross and Self Fertilization in the Vegetable Kingdom.* London: Murray.

DARWIN, C., 1877. *The Different Forms of Flowers on Plants of the same Species.* London: Murray.

DAUMER, K., 1956. Reizmetrische Untersuchung des Farbensehens der Bienen. *Zeitschrift für vergleichende Physiologie, 38:* 413–478.

DAUMDF, K., 1958. Blumenfarben: wie sie die Bienen sehen. *Zeitschrift für vergleichende Physiologie, 41:* 49–110.

DENNIS, B. A. & HAAS, H., 1967. Pollination and seed-setting in diploid and tetraploid red clover (*Trifolium pratense* L.) under Danish conditions. II. Studies of floret morphology in relation to working speed of honey- and bumble-bees (Hymenoptera: Apoidea). *Årsskrift K. Veterinaer-og Landbohøjskole, 1967:* 118–133.

EISNER, T., SILBERGLIED, R. E., ANESHANSLEY, D., CARREL, J. E. & HOWLAND, H. C., 1969. Ultraviolet video-viewing: the television camera as an insect eye. *Science, 166:* 1172–1174.

EISNER, T., EISNER, M., HYYPIO, P., ANESHANSLEY, D. & SILBERGLIED, R. E., 1973a. Plant taxonomy: ultraviolet patterns of flowers visible as fluorescent patterns in pressed herbarium specimens. *Science, 179:* 486–487.

EISNER, T., EISNER, M. & ANESHANSLEY, D. 1973b. Ultraviolet patterns on rear of flowers: basis of disparity of buds and blossoms. *Proceedings of the National Academy of Sciences of the U.S.A., 70:* 1002–1004.

FAEGRI, K. & van der PIJL, L., 1971. *Principles of Pollination Ecology*, 2nd ed. revised. Oxford: Pergamon Press.

FAULKNER, G. J., 1976. Honeybee behaviour as affected by plant height and flower colour in Brussels Sprouts. *Journal of Agriculture Research, 15:* 15–18.

FORSTER, R. E. & LEVIN, M. D., 1967. F$_1$ hybrid muskmelons. II. Bee activity in seed fields. *Journal of the Arizona Academy of Sciences, 4:* 222–225.

FREE, J. B., 1966. The foraging behaviour of bees and its effect on the isolation and speciation of plants. In J. G. Hawkes (Ed.), *Reproductive Biology and Taxonomy of Vascular Plants*. BSBI Conf. Rept. No. 9. Oxford: Pergamon Press.

FREE, J. B., 1970. The effect of flower shape and nectar guides on the foraging behaviour of honeybees. *Behaviour, 23:* 269–286.

FREE, J. B. & SPENCER-BOOTH, Y., 1964. The foraging behaviour of honeybees in an orchard of dwarf apple trees. *Journal of Horticultural Science, 39:* 78–83.

von FRISCH, K., 1960. Über den Farbensinn der Insekten. In *Mechanisms of Colour Discrimination:* 9–40. Oxford: Pergamon Press.

von FRISCH, K., 1965. *Tanzsprache und Orientierung der Bienen.* Berlin, Springer Verlag. *The Dance Language and Orientation of Bees.* L. E. Chadwick, transl. Cambridge, Mass. (1967): Belknap Press, Harvard University.

FROHLICH, M. W., 1976. Appearance of vegetation in ultraviolet light: absorbing flowers, reflecting backgrounds. *Science, 194:* 839–841.

GOLDSMITH, T., 1961. Color vision in insects. In W. D. McElroy & B. Glass (Eds), *Light and Life:* 771–794. Baltimore: Johns Hopkins Press.

GOLDSMITH, T. & BERNARD, G. D., 1974. The visual system on insects. In M. Rockstein (Ed.), *The Physiology of Insects*, 2nd ed., *II:* 165–272. London: Academic Press.

GOPLEN, B. P. & BRANDT, S. A., 1975. Alfalfa flower color associated with differential seed set by leaf cutter bees. *Agronomy Journal, 67:* 804–807.

GRANT, V. A., 1949. Pollination systems as isolating mechanisms in Angiosperms. *Evolution, 3:* 82–97.

GRANT, V. A., 1950. The flower constancy of bees. *Botanical Review, 16:* 379–398.

GRANT, V. A. & GRANT, K., 1968. *Hummingbirds and their Flowers.* New York: Columbia University Press.

GRANT, K., 1966. A hypothesis concerning the prevalence of red coloration in California hummingbird flowers. *American Naturalist, 100:* 85–97.

GULBERG, L. D. & ATSATT, P. R., 1975. Frequency of reflection and absorption of ultraviolet light in flowering plants. *American Midland Naturalist, 93:* 35–43.

HARBORNE, J. B., 1963. Distribution of anthocyanins in higher plants. In T. Swain (Ed.), *Chemical Plant Taxonomy:* 359–388: London: Academic Press.

HARBORNE, J. B., 1967. *Comparative Biochemistry of the Flavonoids.* London: Academic Press.

HARDING, H. G. W., 1950. The colour temperature of light sources. *Proceedings of the Physical Society, 63B:* 685–698.

HENDERSON, S. T., 1970. *Daylight and its Spectrum.* London: Hilger.

HENDERSON, S. T. & HODGKISS, D., 1963. The spectral energy distribution of daylight. *British Journal of Applied Physics, 14:* 125–131.

HENSLOW, G., 1893. *The Origin of Floral Structures through Insect and other Agencies*, 2nd ed. revised. London: Kegan Paul, Trench, Trubner.

HELSOP-HARRISON, J., 1958. Ecological variation and ethological isolation. O. Hedberg (Ed.), *Systematics of Today. Uppsala universitets årsskrift, 1958(6):* 150–158.

HINTON, H. E., 1973a. Natural deception. In R. Gregory (Ed.), *Illusion in Nature and the Arts.* London: Duckworth.

HINTON, H. E., 1973b. Some recent work on the colours of insects and their likely significance. *Proceedings of the British Entomological and Natural History Society, 6* (2): 43–54.

HISDAL, V., 1967. A comparative study of the spectral composition of the zenith sky radiation. *Arbok Norsk polarinstitutt*, Oslo, Serv., 1967: 7–27.

HOCKING, B., 1964. Aspects of insect vision. *Canadian Entomologist, 96:* 320–334.

HÖGLAND, G., HAMDORF, K. & ROSNER, G., 1973a. Trichromatic visual system in an insect and its sensitivity control by blue light. *Journal of Comparative Physiology, 86:* 265–279.

HÖGLAND, G., HAMDORF, K., LANDER, H., PAULSEN, R. & SCHWEMER, J., 1973b. The photopigments in an insect retina. In H. Langer (Ed.), *Photochemistry and Physiology of Visual Pigments:* 164–174. Berlin: Springer Verlag.

HOROVITZ, A., & COHEN, Y., 1972. Ultraviolet reflectance characteristics in flowers of Crucifers. *American Journal of Botany, 59:* 706–713.

HOROVITZ, A. & HARDING, J., 1972. Genetics of *Lupinus*. V, Intraspecific variability for reproductive traits in *Lupinus nanus*. *Botanical Gazette, 133:* 155–165.

ILSE, D., 1928. Uber den Farbensinn der Tagfalker. *Zeitschrift für vergleichende Physiologie, 8:* 658–691.

ILSE, D., 1949. Colour discrimination in the Dronefly, *Eristalis tenax*. *Nature, 163:* 255.

ILSE, D. & VAIDYA, V. G., 1956. Spontaneous feeding response to colours in *Papilio demoleus* L. *Proceedings of the Indian Academy of Sciences, 43:* 23–31.

INGRAM, C., 1967. The phenomenal behaviour of a South African *Gladiolus*. *Journal of the Royal Horticultural Society, 92:* 396–398.

JONES, C. E. & BUCHMANN, S. L., 1975. Ultraviolet floral patterns as functional orientation cues in hymenopterous pollination systems. *Animal Behaviour, 22:* 481–485.

JUDD, D. B., MACADAM, D. L. & WYSZECKI, G., 1964. Spectral distribution of typical daylight as a function of correlated colour temperature. *Journal of the Optical Society of America, 54:* 1031–1040.

KAUFFIELD, N. M. & SORENSEN, E. L., 1971. Interrelations of honeybee preference of alfalfa clones and flower color, aroma, nectar volume, and sugar concentration. *Kansas Agricultural Experimental Station, Research Publication*, No. 163: 14 pp. Manhattan: Kansas.

KAY, Q. O. N., 1976. Preferential pollination of yellow-flowered morphs of *Raphanus raphinastrum* by *Pieris* and *Eristalis* spp., *Nature, 261:* 230–232.

KERNER von MARILAUN, A. & OLIVER, F. W., 1904. *The Natural History of Plants*. London: Gresham.

KEVAN, P. G., 1970. *High Arctic Insect-Flower Relations: the interrelationships of Arthropods and Flowers at Lake Hazen, Ellesmere Island, Northwest Territories, Canada*. Ph.D. thesis, University of Alberta, Edmonton, Canada.

KEVAN, P. G., 1972a. Floral colors in the high arctic with reference to insect flower relation and pollination. *Canadian Journal of Botany, 50:* 2289–2316.

KEVAN, P. G., 1972b. Insect pollination of high arctic flowers. *Journal of Ecology, 60:* 831–847.

KEVAN, P. G., 1972c. Collembola on flowers on Banks Island, N.W.T., *Quaestiones entomologiua, 8:* 121.

KEVAN, P. G., 1973a. Parasitoid wasps as flower visitors in the Canadian high arctic. *Anzeiger für Schädlingskunde, Pflanzen und Umweltschutz., 46:* 3–7.

KEVAN, P. G., 1973b. Flowers, insects, and pollination ecology in the Canadian high arctic. *Polar Record, 16:* 667–674.

KEVAN, P. G., 1976a. Report of activities of Peter G. Kevan as consultant to *RATE*, Meade River and Barrow, Alaska, 2 July to 16 July 1976. Unpubl. Rept. RATE, USACRREL, Hanover, N.H. 7 pp.

KEVAN, P. G., 1976b. Fluorescent Nectar (Technical Comment). *Science, 194:* 341–342.

KEVAN, P. G., GRAINGER, N. D., MULLIGAN, G. A. & ROBERTSON, A. R., 1973. A gray-scale for measuring reflectance and color in the insect and human visual spectra. *Ecology 54:* 924–926.

KEVAN, P. G. & KEVAN, K. K. McE., 1970. Collembola as pollen feeders and flower visitors with observations from the high arctic. *Quaestiones entomologiua, 6:* 311–326.

KNOLL, F., 1921. Insekten und Blumen. Experimentale Arbeiten zur Vertiefung unserer Kenntnisse Über die Wechselbeziehungen zwischen Pflanzen und Tieren. II. *Bombylius fuliginosus* und die Farbe der Blumen. *Abhandlungen der Zoologisch-botanischen Gesellschaft in Wien, 12:* 17–119.

KNOLL, F., 1922. *ibidem* III. Lichsinn und Blumenbesuch der Falters von *Macroglossum stellatarum*. *Abhandlungen der Zoologisch-botanischen Gesellschaft in Wien, 12:* 127–378.

KRONFELD, M., 1889. Uber die biologsichen Verhaltnisse der Aconitumblüte. *Botanische Jahrbücher für Systematik Pflanzengeschichte und Pflanzengeographie, 11:* 1–20.

KUGLER, H., 1930–1932. Blütenökologie Untersuchungen mit Hummeln. I, III, IV. *Planta, 10:* 229–280; *16:* 227–276; 534–553.

KUGLER, H., 1936. Die Ausnutzung der Saftmalsumfärbung bie den Roszkastanienblüten durch Bienen und Hummeln. *Bericht der Deutschen botanischen Gesellschaft, 60:* 128–134.

KUGLER, H., 1938. Sind *Veronica chamaedrys* L. und *Circaea lutetiana* L. Schwebefliegenblumen? *Botanische Archiv., 39:* 147–165.

KUGLER, H., 1943. Hummeln als Blütenbesucher. *Ergebnisse der Biologie, 19:* 143–323.

KUGLER, H., 1950. Der Blütenbesuch der Schammfliege (*Eristalomyai tenax*) *Zeitschrift für vergleichende Physiologie, 32:* 328–347.

KUGLER, H., 1951. Blütenökologische Untersuchungen mit Goldfliegen (Lucilien). *Bericht der Deutschen botanischen Gesellschaft, 64:* 327–341.

KUGLER, H., 1956. Über die optische Wirkung von Fliegenblumen auf Fliegen. *Bericht der Deutschen botanischen Gesellschaft, 69:* 387–398.

KUGLER, H., 1963. Untersuchungen auf Blüten und ihr Zustandekammen. *Planta, 59:* 296–329.

KUGLER, H., 1970. *Blütenökologie.* Stuttgart: Fischer Verlag.

KUHN, A., 1927. Über den Farbensinn der Bienen. *Zeitschrift für vergleichende Physiologie, 5:* 762–800.

KULLENBERG, B., 1956. On the scents and colors of *Ophrys* flowers and their specific pollinators among the aculeate Hymenoptera. *Svenska botanisk tidskrift, 50:* 25–46.

KULLENBERG, B., 1961. Studies in *Ophrys* pollination. *Zoologiska bidrag fran Uppsala, 34:* 1–340.

KULLENBERG, B., 1973. New observations on pollination of *Ophrys* L. (Orchidaceae). *Zoon (Suppl.)* No 1. 9–14.

KULLENBERG, B. & BERGSTROM, G., 1976. The pollination of *Ophrys* orchids. *Botaniska notiser, 129:* 11–20.

LAND, E. H., 1977. The retinex theory of color vision. *Scientific American, 237* (6), 108–128.

LELEJI, O. I., 1973. Apparent preference by bees for different flower colours in cowpeas (*Vigna sinensis* (L.) (Savi ex Hassk.)). *Euphytica, 22:* 150–153.

LEVIN, D. A., 1969. The effect of corolla color and outline on interspecific pollen flow in Phlox. *Evolution, 23:* 444–455.

LEVIN, D. A., 1972. The adaptiveness of corolla color variants in experimental and natural populations of *Phlox drummondii. American Naturalist, 106:* 57–70.

LEVIN, D. A. & SCHAAL, B. A., 1970. Corolla color as an inhibitor of inter-specific hybridization in *Phlox. American Naturalist, 104:* 273–283.

LEWIS, D., 1965. The genetic integration of breeding systems. In J. G. Hawkes (Ed.), *Reproductive Biology and Taxonomy of Vascular Plants.* BSBI. Conf. Rept. No.9, 20–25 + plate. Oxford: Pergamon Press.

LEX, T., 1954. Duftmale an Blüten. *Zeitschrift für vergleichende Physiologie, 36:* 212–234.

LIST, R. J. (Ed.), 1968. *Smithsonian Meteorological Tables, 6th rev. ed., 1966.* Washington: Smithsonian Institution Press.

LOTHMAR, R., 1933. Neue Untersuchungen über den Farbensinn der Bienen, mit besonderer Berücksichtigung des Ultravioletts. *Zeitschrift für vergleichende Physiologie, 19:* 673–723.

LUBBOCK, J., 1881. Observations on ants, bees, and wasps. Part 9. Colours of flowers as an attraction to bees: experiments and considerations thereon. *Journal of the Linnean Society of London (Zoology), 16:* 110–112.

LUTZ, F. E., 1924. Apparently non-selective characters and combinations of characters, including a study of ultraviolet in relation to flower visiting habits of insects. *Annals of the New York Academy of Sciences, 29:* 181–283.

LUTZ, F. E., 1933. "Invisible" colors of flowers and butterflies. *Natural History, 33:* 565–576.

MACIOR, L. W., 1968. Pollination adaption in *Pedicularis groenlandia. American Journal of Botany, 55:* 927–932.

MACIOR, L. W., 1975. The pollination of *Delphinium tricorne* (Ranunculaceae). *American Journal of Botany, 62:* 1009–1016.

MACNICHOL, E. F., 1964. Three pigment color vision. *Scientific American,* December 1964.

MANNING, A., 1956a. The effect of honey guides. *Behaviour, 9:* 114–139.

MANNING, A., 1956b. Some aspects of the foraging behaviour of bumblebees. *Behaviour, 9:* 164–201.

MATHER, K., 1947. Species crosses in *Antirrhinum.* I. Genetic isolation of the species *majus, glutinosum,* and *orontium. Heredity, 1:* 175–183.

MAZOKHIN-PORSHNYAKOV, G. A., 1959. Otrazheniye ultrafioletovikh luchei tsvetkami rastenii i zreniye nasekomikh. *Entomologicheskoe obozrenie, 38:* 321–325.

MAZOKHIN-PORSHNYAKOV, G. A., 1962. Kolorimetricheskoye dokazatel'stvo trikhromazii tsvetovogo zreniya pchelinikh (na primere shmelei). *Biofizika 7:* 211–217.

MAZOKHIN-PORSHNYAKOV, G. A., 1966. Recognition of coloured objects by insects. *Proc. Int. Symp. on the Functional Organization of the Compound Eye:* 163–170. Stockholm, October 25–27, 1965. Oxford: Pergamon Press.

MAZOKHIN-PORSHNYAKOV, G. A., 1969. *Insect Vision.* (R. & L. Masironi, trans. T. H. Goldsmith (Ed.).) New York: Plenum Press.

MENZEL, R., 1971. Über den Farbensinn von *Paravespula germanica* F. (Hymenoptera): ERG und selektive Adaptation. *Zeitschrift für vergleichende Physiologie, 75:* 96–104.

MOFFETT, J. O., SAITH, L. S., BURKHARDT, G. G. & SHIPMAN, C. W., 1975. Influence of cotton genotypes on floral visits of honeybees. *Crop Science, 15:* 782–784.

MOGFORD, D. J., 1974. Flower colour polymorphism in *Cirsium palustre. Heredity, 33:* 257–263.

MÜLLER, H., 1876. On the relation between flowers and insects. *Nature, 15:* 178–180.

MULLIGAN, G. A., 1972. Autogamy, allogamy, and pollination in some Canadian weeds. *Canadian Journal of Botany, 50:* 1767–1771.

MULLIGAN, G. A. & FINDLAY, J., 1970. Reproductive systems and colonization in Canadian weeds. *Canadian Journal of Botany, 48:* 859–860.

MULLIGAN, G. A. & KEVAN, P. G., 1973. Color, brightness, and other floral characteristics attracting insects to the blossoms of some Canadian weeds. *Canadian Journal of Botany, 51:* 1939–1952.

NEWTON, I., 1704. *Optiks: or, a Treatise of the Reflexions, Refractions, Inflexions and Colours of Light.* London: The Royal Society.

ORNDUFF, R. & MOSQUIN, T., 1970. Variation in the spectral qualities of flowers in the *Nymphoides indica* complex (Menyanthaceae) and its possible adaptive significance. *Canadian Journal of Botany, 48:* 603–605.

PANFILOV, D. V., SHAMURIN, V. F. & YURTSEV, B. A., 1960. O soprezhennom rasprostanenii shmeleii babovikh v arktikye. *Byulleten' Moskovskogo obshchestva ispastatelei priroda, 65:* No. 3.

PETERSEN, B., 1968. *Pollination of some tundra plants with minature flowers on Niwot Ridge in Boulder County, Colorado.* Ph.D. thesis, University of Colorado, Boulder.

PREISNER, E., 1973. Reaktionen von Reichrezeptoren männlicher Solitarbienen (Hymenoptera: Apoidea) auf Inhaltsstoffe von *Ophrys*-Blüten. *Zoon, (Suppl.), 1:* 43–54.

PROCTOR, M. C. F. & YEO, P., 1973. *The Pollination of Flowers.* London: Collins.

RAVEN, P., 1972. Why are bird-visited flowers predominantly red? *Evolution, 26:* 674.

RICHTMYER, F. K., 1923. The reflection of ultraviolet by flowers. *Journal of the Optical Society of America, 7:* 151–168.

RIPPS, H. & WEALE, R. A., 1969. Color vision. *Annual Review of Psychology, 20:* 193–216.

SCORA, R. W., 1964. Dependency of pollination on patterns in *Monarda* (Labiateae). *Nature, 204:* 1011–1012.

SHAMURIN, V. F. & TIKHMENEV, E. A., 1974. Vzaimosvyazi mezhdu entomofil'nimi rasteniyami i antofil' nimi v biogeotsenozakh arktiki. *Zhurnal obshchei biologii, 35:* 243–250.

SILBERGLEID, R. E., 1976. Visualization and recording of long wave ultraviolet reflection from natural objects. Parts 1 and 2. *Functional Photography, 11:* 20–29, 30–33.

SPRENGEL, C. K., 1793. *Das endeckte Geheimniss der Natur im Bau und in der Befruchtung der Blumen.* Berlin: Vieweg. (Facsimile Drucken, Wissenschaftliche Classiker, Vol. III, 1898. Mayer u. Müller, Berlin).

STRUWE, G., 1972a. Spectral sensitivity of the compound eye in butterflies (*Heliconius*). *Journal of Comparative Physiology, 79:* 191–196.

STRUWE, G., 1972b. Spectral sensitivity of single photoreceptors in the compound eye of a tropical butterfly (*Heliconius numata*). *Journal of Comparative Physiology, 79:* 197–201.

THIEN, L. B., 1971. Orchids viewed with ultraviolet light. *Bulletin of the American Orchid Society, 10:* 877–880.

THOMAS, I. & AUTRUM, H., 1965. Die Empfindlichkeit der dunkel- und hell-adaptierten Biene (*Apis mellifica*) für spekrale Farben; zum Purkinje-Phänomen der Insekten. *Zeitschrift für vergleichende Physiologie, 51:* 204–218.

THOMPSON, W. R., MEINWALD, J., ANESHANSLEY, D. & EISNER, T., 1972. Flavonols: pigments responsible for ultraviolet absorption in nectar guide of flowers. *Science, 177:* 528–530.

THORP, R. N., BRIGGS, D. L., ESTES, J. R. & ERIKSON, E. H., 1975. Nectar fluorescence under ultraviolet irradiation. *Science, 189:* 476–478.

THORP, R. N., BRIGGS, D. L., ESTES, J. R. & ERIKSON, E. H., 1976. (Reply to Kevan (1976b).) *Science, 194:* 342.

TINBERGEN, N., MEEUSE, B. J. D., BOEREMA, L. K. & VAROSSIEAU, W. W., 1942. Die Balz des Samfalters *Eumenis* (*Satyrus*) *semele* L. *Zeitschrift für Tierpsychologie, 5.*

UTECH, F. H. & KAWANO, 1975. Spectral polymorphisms in angiosperm flowers determined by differential ultraviolet reflectance. *Botanical Magazine, 88:* 9–30.

VALENTINE, D. H., 1975. The taxonomic treatment of polymorphic variation. *Watsonia, 10:* 385–390.

WATT, W. B., HOCH, P. C. & MILLS, S. G., 1974. Nectar resource use by *Colias* butterflies; chemical and visual aspects. *Oecologia, 14:* 353–374.

WEEVERS, T., 1952. Flower colours and their frequency. *Acta botanica neerlandica, 1:* 81–92.

WICKLER, W., 1968. *Mimicry in Plants and Animals.* (R. D. Martin, transl.) London: Weidenfeld and Nicolson.

WILLIAMS, N. H. & DODSON, C. H., 1972. Selective attraction of male euglossine bees to orchid fragrances and its importance in long distance pollen flow. *Evolution, 26:* 84–95.

WINCH, G. T., BOSHOFF, M. C., KOK, C. J. & DUTOIT, A. G., 1966. Spectral radiometric and colorimetric characteristics of daylight in the southern hemisphere: Pretoria, South Africa. *Journal of the Optical Society of America, 56:* 456–464.

WOODSON, R. E., 1964. The geography of flower color in Butterflyweed. *Evolution, 18:* 143–163.

WRIGHT, W. D., 1969. *The Measurement of Colour*, 4th ed. London: Hilger.

WYSZECKI, G. & STILES, W. S., 1967. *Color Science: Concepts and Methods, Quantitative Data and Formulas.* New York: Wiley.

YAZAKI, Y., 1976. Co-pigmentation and color change with age in petals of *Fuschia hybrida*. *Botanical Magazine, 89:* 45–57.

EXPLANATION OF PLATES
PLATE 1 (*see Frontispiece*)

Floral colours as humans see them (right side of flowers), and as insects (bees) might see them (left side of flowers), as suggested by behavioural and physiological experiments on insect colour vision, and by colorimetry (colour mixing) according to the three insect primary colours analogised with the three human primary colours.

A. *Epilobium latifolium* and *E. angustifolium*. B. *Dryas integrifolia*. C. *Arnica alpina*. D. *Erigeron compositus* (also many other white rayed and yellow disced composites).

PLATES 2 to 5

Photographs showing spectral analyses of floral coloration of alpine plants of the Colorado Rocky Mountains (*c.* 3500 m, Pennsylvania Mt., near Fairplay, Colorado). White lines provide examples of matching reflectances of floral parts with chips on a calibrated gray-scale. By using 7 broad-band monochromatic filters a spectral reflectance curve can be generated (centre graph). Results from 4 filters are shown representing the primary colours in the IVS (3 photographs enclosed by solid line) and in the HVS (3 photographs enclosed by dashed line): 18A (upper left) is UV or Insect-Blue, 98 (upper right) is Blue or Insect-Green, 61 (lower right) is Yellow-Green (a compromse between Green and Yellow or Insect-Red), and 25 (lower left) is Red or Insect-Infrared. Using proportionate reflectances in the primary colour wavebands, trichromaticity plots can be made for the IVS (left) and HVS (right). Vegetation colours were also analysed and show as a relatively dull background against which the blooms contrast.

(Photographs, developing, and printing were by David Davidson, Biology Department, University of Colorado, Colorado Springs from the summer of 1977.)

PLATE 2

Zygadenus elegans (Liliaceae) on right and *Cardamine cordifolia* (Cruciferae) on left.

Of the two white flowers, *C. cordifolia* is brighter and whiter, whereas *Z. elegans* is duller and slightly yellower. Within the corolla of *Z. elegans* overall reflectance is higher than from the outside and there are splotches with associated nectaries which are yellow to green (herein yellow). Reflectance patterns from the area of these splotches are complex and not fully analysed.

Curve and point numbers: 1, 2, 3: *C. cordifolia*: 1, corolla; 2, calyx; 3, vegetation. 4–7; *Z. elegans:* 4, inside of corolla; 5, outside of corolla; 6, splotches within corolla; 7, vegetation.

PLATE 3

Campanula rotundifolia (Campanulaceae).

A blue flower showing an ultraviolet absorbing centre, including the ovary, within the corolla. Otherwise the flower, inside and out, is uniformly colored in both IVS and HVS.

Curve and point numbers: 1, outside of corolla; 2, inside of corolla; 3, centre of inside of corolla; 4, ovary; 5, vegetation.

PLATE 4

Solidago spathulata (Compositae).

Three age classes of these yellow inflorescences are shown: Left, young inflorescences in which the disc florets have not yet opened; Centre, old inflorescences in which the disc florets have turned brownish and the ray florets become somewhat bleached; and Right, inflorescences presenting pollen and receptive stigmata.

In the HVS little difference can be seen in the coloration of the age groups, the rays of all being pale yellow, and the discs of the young and receptive inflorescences a more intense yellow. The discs of the old inflorescences show a position near yellow, but their low reflectance makes them brownish.

In the IVS the three age classes show clear differences in coloration, especially of the ray florets. As the inflorescences change from young to receptive, more ultraviolet is reflected; this diminishes again as the florets age. The disc florets, meanwhile, remain an intense insect-red from young to receptive, and shift to a duller and less saturated colour with age.

Curve and point numbers: 1, 2: new inflorescences: 1, ray florets; 2, disc florets. 3, 4: receptive inflorescences: 3, ray florets; 4, disc florets. 5, 6: old inflorescences: 5, ray florets; 6, disc florets. 7, vegetation.

PLATE 5

Pedicularis groenlandicus (Scrophulariaceae).

In this generally pinky mauve colored flower, the patterns of coloration are complex. The 'forehead' is white in the HVS, but pale yellow in the IVS. The 'trunk' and 'ears' are pinky mauve in the HVS, but pale and dull yellowish green in the IVS with the ears distinctly greener. The upper flowers, which are not visited by pollinating bumblebees, and are not fully open, are generally much duller in reflectance across both HVS and IVS, although they are similarly coloured. The 'face' of the elephant's head is Insect-Black with almost no reflection (4% or less) in all IVS wavebands, but it is dark red or magenta (*vide* Macior 1968) in the HVS.

Curve and point numbers: 1, 'forehead'; 2, 'trunk'; 3, 'ears'; 4, 'face'; 5, upper flowers (general reflectance); 6, vegetation.

PLATE 2

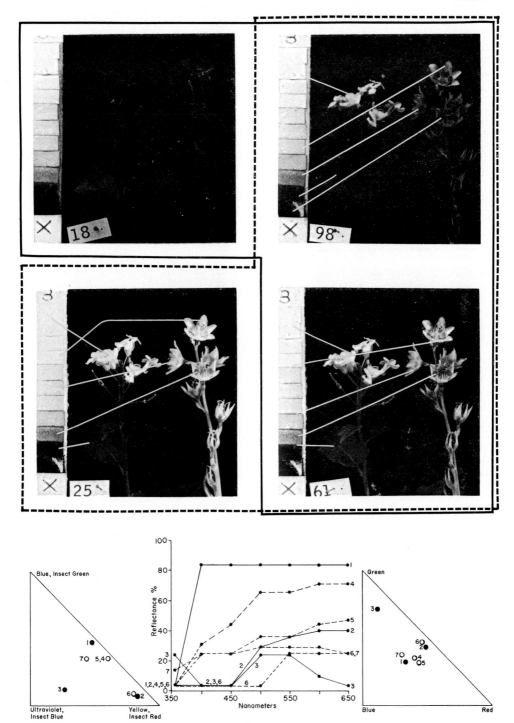

PLATE 3

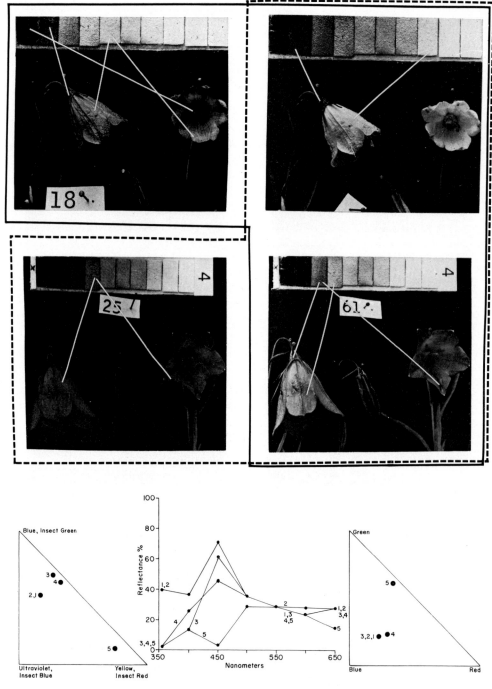

P. G. KEVAN

PLATE 4

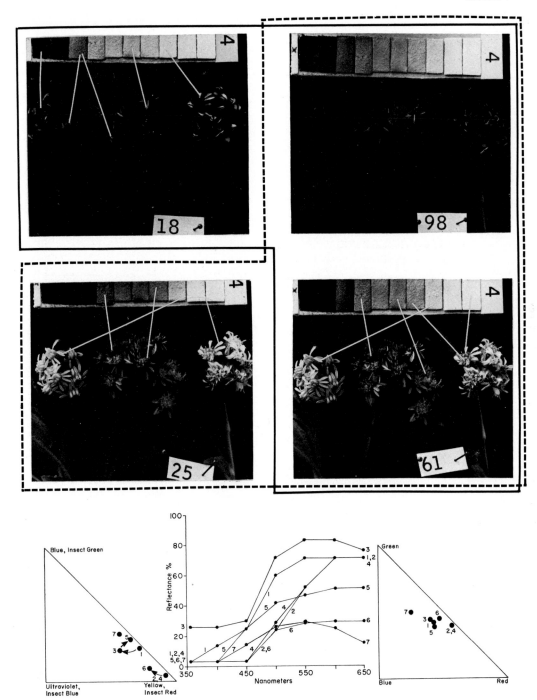

P. G. KEVAN

PLATE 5

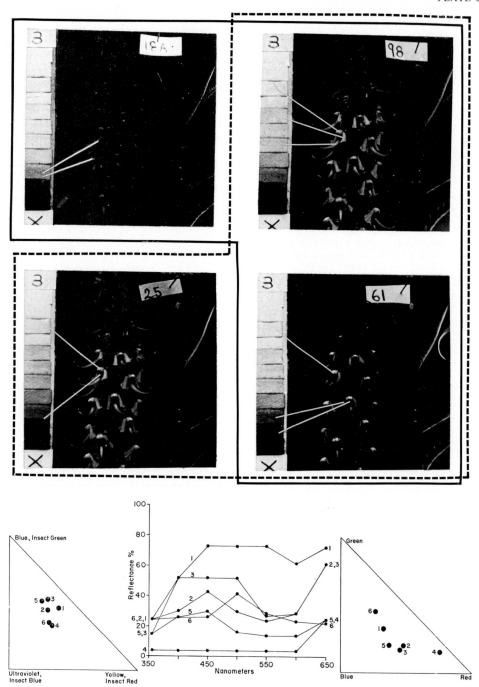

P. G. KEVAN

Reproductive integration and sexual disharmony in floral functions

L. van der PIJL

236 Sportlaan, The Hague, The Netherlands

The original flower is considered as being perfect, constituted through a forced marriage between its two main functions (the issue and reception of pollen), as long as pollen remains the main attractant. However, secondary adaptations, for instance to dicliny or seed dispersal, can upset this balance.

It is envisaged that pollen presentation was originally diffuse, adapted to diffuse primitive visitors. More localised presentation led to the development of specialised structures which influenced other floral functions. Increased specialisation also caused disharmonies between functions, encouraging the development of dicliny, or a division of labour within the inflorescence. Pollination is merely one of many floral functions which may interact, and increased specialisation has caused problems in the success of floral functions, as for instance between pollination and fertilisation.

KEY WORDS:—pollination—fertilisation—Orchidaceae—Compositae—origin of angiovuly—dicliny—nectaries.

CONTENTS

FUNCTIONS

Let it be said initially that the use of the term 'sex' for flowers is not quite relevant, unless the production of two kinds of spore can be considered as an activity of the sexual cycle. The direct floral functions will be considered here, without concentrating solely on the promotion of cross-pollination for a remote advantage, a common pre-occupation of floral ecologists. They should not ignore the final (and more ancestral) functions of the flower.

We have to consider progressive changes in floral structure in relation to changes in visitors, but we must not ignore features of internal engineering involved in the physiological integration of floral functions. At the same time, our considerations of functions do not onesidedly support the purely typological 'Bautypen' of some authors, who tend to ignore ecological influences in the structural classes of flowers, relying mainly on orthogenesis to explain floral morphology.

The main functions of flowers are:

(a) Formation and release of microspores, with presentation to a vector.
(b) Reception of pollen and guidance of male gametophytes.
(c) Gamete fusion.
(d) Formation of seeds, and associated dispersal structures.
(e) Auxiliary functions for the attraction or exclusion of animal visitors.

Interaction between these functions generally acts forewards in time, providing a residual influence in later development of organs (e.g. seed, fruit). However, it is frequently found that organs of ancient origin, latterly suppressed in expression, may become re-expressed, fulfilling functions earlier in the developmental sequence. This provides developmental economy. Examples of such interactions backwards in time include functions of dispersal in the seed or fruit influencing organs for pollination or fertilisation. Such premature expression of fruit dispersal genes can also interact with genes expressed early but with residual influences. Thus, the situation becomes very complex and difficult to interpret clearly. However, in general, organs which function late in floral ontogeny have been 'enveloped' by outer organs of more recent origin, and these former are the most ancient in origin. Older and newer functions play 'Bo-Peep' in the flower, the older being more central.

The angiovulate flower, requiring precision in pollen reception, is considered as originally entomophilous (rightly in my view), as it seems unlikely that angio-ovuly would have arisen in response to anemophily. This original flower was probably hermaphrodite, capable of both issue and reception of pollen. This system seems to have been necessary as long as attractants were confined to pollen (for beetles) as now in *Drimys* spp. Pollen had obtained an ecological, collateral function (by no means new in angiovulates) as food. Its attractivity was through smell (also not new); pollen formed the earliest floral odour. Surrogate solid food bodies, and sapromyophilous deceit of animals complemented or replaced pollen odour at an early stage, although it is likely that floral perfumes, attractive to Hymenoptera, developed much later.

Other early cases of attraction were those where ovipositing parasites visit the flower for two separate functions, pollen (food) and egg-laying. This is possible even in diclinous plants, and perhaps in the preangiovulate condition. To satisfy colleagues who have considered such primary beetle-pollination as a mere speculation (having experience only of later, more highly adapted flower-beetles) I refer to the devices of secondary pollen-flowers treated below.

Floral parasitism is likely to result in rapidly evolving specificity between the parasite and host (pollinator and plant). Such parasitism is still evident in some Cycads, and perhaps occurred in early Cycadeoidea. There the enclosed, mixed, protandrous cones could be visited by parasitic beetles twice, once for pollen (food) and once for egg-laying, resulting in pollination and thus entomophily. Analogues of such balanced parasitism are found today in highly derived angiosperms (Annonaceae, Palmae, Araceae, Moraceae), which can be considered as secondary pollen-flowers.

In early angiosperms, pollen presentation may originally have been diffuse on laminar stamens, for diffuse, crawling, internal visitors. The stamens may have required strengthening as a protection from gnawing. Better presentation to external deliberate pollen-collectors is already visible in the primitive Winteraceae, possessing 'conventional' stamens. However, secondarily flattened stamens occur in more derived families. When nectar was later employed (and Hymenoptera participated), it could complement and replace pollen as an attractant and food, and separation of the floral functions of pollen issue and reception was thus facilitated.

With further development, microsporophylls remained rather uniform although the functions of pollen presentation for food (attraction) and dispersal were sometimes separated. Refinements in pollen presentation evolved, often resulting in economic usage (e.g. pollen release through visitor activity). In a highly derived group, the orchids abandoned the use of pollen as a food attractant, and with daring precision combined numerous grains in one unit (pollinium).

Pollen reception on more complicated pistils led to refinements in efficiency and centralisation. Thus, a number of open stigmatic crests (with limited sieving) developed to special stigmas on loose carpels (which still require diffuse pollinator activity of a non-specialised kind), to loose carpels with united stigmas, and finally to centralised syncarpy. Further co-ordination of issue and reception of pollen led ultimately to precision in united gynostemium (Asclepiadaceae). Here is a clear example of a developmental sequence showing that increased specialisation of pollen reception is having a marked forwards influence on fruit formation and dispersal.

The 'semaphylls' coevolved with gynoecium centralisation to aid visitor specificity through specialised attraction and guidance, providing more integration with the outer world.

POTENTIAL DISHARMONIES BETWEEN FUNCTIONS OF FLORAL FEATURES, LEADING TO DICLINY

It seems remarkable that the androecium and gynoecium can live together harmoniously in a flower. The stamens are laterally placed, and short-lived. The megasporophylls (carpel, ovary), with tightly enveloped sporangia (ovules) are apically placed, destined for a long life, culminating in the fruit, and accumulating food. The necessary acropetal changes required to express these differences, in gene activation, in hormone balance, and in vascularisation are imposing. Physiologists work on 'sex-expression', and floral ecologists should not neglect these investigations of morphogenetic interactions. Some cases of dicliny, especially those which are environmentally labile, may be explicable through such work (for instance male sterility through disharmony in male meiosis).

It has been remarked that the invariable sequence of floral organs (with the persistent female ones on top) may promote necessary vascularisation for the latter. This sequence has no roots in primitive ontogeny; it was rarely found in preangiospermous strobili, and is not automatically repeated in later monoecious pseudanthia as will be shown for the Compositae.

Ecologically speaking, we must remind ourselves that seed-dispersal not only uses and transforms floral parts after seed-formation, but earlier anticipates the requirements of seeds and fruits. In my book (van der Pijl, 1972), documentation of this interaction is presented, especially for flower position, ovule vascularisation and ovule number. Another example is shown by seeds which lack a testa for ecological reasons, such as parasites and viviparous seeds; this can already be expressed in the flower through reduced or missing ovule integuments. Specialised awns for burrowing and dispersal are already differentiated in the grass flower before anthesis. Lack of co-ordination between the flower and fruit often results in adaptive sacrifices. Thus the mimosoid head of a *Parkia* produces only one or two very large seed pods from hundreds of minute flowers.

To mediate between pollen release and pollen reception, a vector is required, and this is not often the same as that involved in seed dispersal (even in grasses). In this group, a potential disharmony exists between the anemophily of pollination, and seed-dispersal through the shedding of the indehiscent fruits (from exposed, anemophilous, monovulate ovaries) to a water surface on the soil.

Seed dispersal by animals is older, and started with higher life forms than pollen dispersal by animals. In my book on dispersal (van de Pijl, 1972), I presented arguments to support the thesis that angiovuly, which allowed more accurate and economic pollination (unfortunately named angiospermy) disturbed seed-dispersal syndromes of an earlier origin.

An example of feedback from carpal adaptations leading to dicliny ("uni-sexuality") through sexual disharmony concerns the precocious enlargements of the ovarial wing of *Acer*, which is directed by genes expressing adaptations concerned with fruit dispersal. This development is autonomous even of fertilisation, and seems to have led to alternatives of androecy or gynoecy, the production of stamens being structurally impracticable in a gynoecious flower. In a similar manner, dicliny may be associated with differently advantageous positions of the flower for pollen issue as against fruit-dispersal.

This shift from reception to presentation is clear in *Eupatorium*, *Gynura* and that can lead to conflict and disharmony. The evolution of crowded narrow sympetalous florets with weakly developed filaments produced a new condition, forming a substrate for novel integration, specialisation and relocation of functions. A number of the tubiflorous Compositae, usually regarded as primitive, have two very long stigmatic branches which have largely lost their original function. Not only do they sweep pollen out of the floret (considered normal in the family), but present it to collectors well clear of the capitulum surface. This fact has been ignored in textbooks, and even in a classic paper by Small (1915) on pollen presentation. This compensative function of presentation is especially evident when the styles shorten again after presentation. This secondary pollen presentation seems to be a continuation of the condition in some Campanulaceae and Rubiaceae, which may be ancestral to the Compositae. In general, it only lasts through the first phase of anthesis.

This shift from reception to presentation is clear in *Eupatorium*, *Gynura* and *Ageratum*, where the coloured stigmas also assume the role of visual attraction, and deliberate pollen collection from the style branches occurs. Similar combinations of functions are found in disk florets of groups in which the ray-florets have lost the capacity to present pollen, and specialise in reception, this affording a division of labour. The stigma branches are often temporally combined into a spiked club with an inner stigmatic space. In *Solidago*, the two stigma branches are even fused apically, leaving a narrow basal lantern-slit open underneath (Plate 10). At first sight this seems incapable of fulfilling a female function, but the margins of the slit become pollen-receptive late in anthesis. In *Tussilago*, the disk florets are similar to those in *Solidago*, but here the original stigmatic function is suppressed by the new one. One could call it sex reversal of the style into a new stamen. In the apparently normal ovary, the ovules abort. In the Melampodineae and Calenduleae similar 'male' disk florets occur, apparently independently. A series of stages showing the gradual reduction of the ovary can be demonstrated, while in some the nectary also aborts. The development of the ovule should be checked when investigating the floral biology of these plants, especially in *Centaurea* in which female sterility occurs in certain florets independently of structural changes, and can pass unsuspected.

Sexual differentiation inside the capitulum seems not to be just internal specialisation. It may also act in an external ecological strategy for cross-pollination. The original protandry (in general effective for this purpose in a raceme) remained in small heads with few synchronous florets, all in one phase. It lost this effect in a larger, flat disk with successive anthesis, where central (apical) alighting of insects occurs. It was replaced by protogyny of the compound inflorescence as a whole, with female flowers at the rim opening first.

We will now consider further the divorce of the two main functions of the flower, pollen issue and reception, through dicliny. Dicliny apparently arose in diverse groups in diverse ways, and is secondary and often ephemeral; sometimes via heteromorphy, and sometimes from self-incompatible plants (already outbreeders) which leads to complications. Sometimes sexuality proceeds in diploids so far that 'male' and 'female' plants differ vegetatively, showing secondary sexual characteristics, as do most animals. This may fit them to occupy different ecological niches (Darwin, 1877), thus minimising competition between the sexes which may have differing energy loads at different seasons. In such a manner, the sex-ratio may also be adjusted to maximum reproductive efficiency. On the other hand, such niche specialisation in organisms with limited motility may create something of a barrier to pollination. The colonisation of islands is often mentioned as a steering selective force behind the origin of dicliny, but it is not clear to me why this should be so, and the predominance of dicliny in New Zealand, often apparently of local autonomous origin, is an unsolved mystery.

Monoecious dicliny (with a physiological basis) arose even in the orchids, where the functional bond between the highly specialised androecium and gynoecium would seem unbreakable. This happens when the superabundance of floral drugs induces male Euglossidae to return repeatedly to the same flower for perfume. Here a self-incompatible bisexual flower would result in much wastage of pollen. These cases also illustrate an unbalance in issue and reception of pollen, in that issue by downward explosion cannot be countered by upward redisposition onto the style (rostellum cavity). Thus the evolution of a differently constructed and positioned female flower became advantageous in these cases of overprecise pollination.

A similar separation of sexual function towards dicliny tends to occur at the other extreme, very imprecise pollination, as in anemophily. The requirements of good pollen release and good reception in anemophiles are too disparate for harmonious coexistence : downwards release from pendulous, shaking stamens, and reception by a vertically positioned large erect free stigma. Of course, reasoning on such functional grounds must be considered to be complementary to arguments concerning the breeding system, which rightly emphasise the danger of selfing in anemophiles.

It appears simplistic (and unecological) to ascribe the production of large quantities of pollen, or unisexuality, or the reduction of perianth or ovule number, wholly to original anemophily. This was recently done by so critical a botanist as Stebbins (1974) for instance in the Araceae and Palmae, which are both basically entomophilous (cantharophilous). The production of odour, the nature of the pollen and its release are far from anemophilous in the Araceae. Neither do their carpels show oligovuly. Dicliny in these groups is not primitive, but is a derived phenomenon, facilitating, in some genera, second order protogyny between flower specialists in an integrated, monoecious trap. In a similar way, tropical, entomophilous Fagaceae also show dicliny and monovuly, preadaptive to their (later) anemophily. Sexual aggregation here ensures the pollination of the unattractive pistillate flowers.

Darwin (1877) could not accept that a conversion to dicliny could have been affected solely in response to requirements for better cross-pollination, and invoked unfavourable external conditions when exploring the connection between dicliny and preceding dichogamy. He considered the development of dicliny to be causal, relying on the differential influence on the separate organs of growth and nutrition, and of later seed formation. We may extend this view to other functions. The harmony which balances release and reception of pollen may also be disturbed by changes in syndromes and degrees of attractivity that occur during anthesis, for instance, in strongly dichogamous flowers in which attractivity (colour, odour) ceases on the second day and this may encourage the development of dicliny. In cases where the production of pollen and/or nectar also changes, the energy budget of the pollinator is also affected. Examples are *Aesculus* (protogyny with dicliny), *Echium* (protandry, often with gynodioecy) and *Silene* (with increasing dioecy). Protandrous Dipsacaceae showing gynodioecy deserve investigation along these lines.

Such unbalance may also occur when protandrous flowers offer only pollen on the first day, and only nectar on the second day, especially when a shift to obligate nectar feeders (as many Lepidoptera) occurs. Such unbalance is possible in *Tilia* and in single flowers of some Labiatae, although it seems that those pollinated by Lepidoptera offer nectar throughout. In Labiatae, lacking self-incompatibility, protandry is so strongly expressed that it may lead to loss of attractivity in the last, receptive, stage. Research is planned which will investigate whether gynodioecy, frequent in this group, occurs in response to this disharmony. In certain Labiatae, which show explosive pollen release, the second phase of the flower becomes a quite different type. Gynodioecy in the Labiatae seems in no way to represent a transition phase towards full dioecy, as it appears to be in many groups (some antipodean *Fuchsia*), where bisexual individuals show inferior seed production. There the term subdioecy is applied for gynodioecy, and this may be a useful distinction.

Suggestions that gynodioecy merely serves as a means to increase the ratio of ovules to pollen is not yet proven, nor is the thesis that gynodioecy acts to regulate and optimise genetically determined sex ratios.

The opposite phenomenon, entomophilous protogyny, has been considered to have caused disharmony leading to dicliny in the case of sympetalous *Plantago* spp. (pollen flowers) switching partially to anemophily. It may be that the duration of the receptive first phase in bisexual flowers proved too short for wind pollination to be effective in regions where anemophily prevails. Pollination of the nectarless flowers by pollen-collecting syrphids has been shown to be more effective than anemophily in the Netherlands (Stelleman, 1978). However, in the Netherlands, gynodioecy in *Plantago* is very rare, in contrast to other regions, perhaps because syrphids would be unlikely to visit separated females in this area (gynodioecy in the Labiatae shows similar geographical trends). Work is currently in progress to determine whether in females of *Plantago* (with larger and better nourished gynoecia than in bisexual plants), the anemophilous reception is better and of longer duration, thus favouring dicliny.

The phenomenon of androdioecy reflects special features of pollen issue. As pollen issue is less complex than pollen reception, it is understandable that androdioecy is so uncommon (as noted by Darwin). In some cases of androdioecy that are quoted (for instance *Sterculia* and *Vitis*), it may be that apparently bisexual individuals are in fact functional females, there being full dioecy. Dioecy is often not evident from external morphology, and it is necessary to investigate pollen fertility in order to establish androdioecy. Androdioecy has been considered to serve a regulative function in cases where pollination is handicapped, for instance, by too small a stigma (relict from earlier syndromes), or when the pollen load is diffusely placed on the vector.

Andromonoecy occurs in many grasses, often producing male (sterile) spikelets accompanying caryopses (*Lamarckia*). In *Spinifex squarrosus* (well known for its rolling wind-balls), andromonoecy has progressed to androdioecy, where it is also apparently associated with seed-dispersal, all-important in grasses. The synapto-spermous inflorescence balls reduce the number of seed-bearing spikelets to encourage dispersal efficiency, and at the same time pollen donation is improved through the many-flowered male spikelets.

INTERACTION THROUGH NECTARIES

We have already seen that special means of attraction may have unbalancing effects on various floral functions; for example, temporally limited nectar supply may encourage dicliny. In detail, we will pass over parameters of nectar quality, and quantity, the latter in particular being finely adjusted to neither exceed nor underpay specific requirements of the pollinator.

The location of nectaries may inhibit other adaptations. Thus in a group with carpel secretion, male flowers lack nectar. In *Rhamnus* spp. male flowers have acquired novel nectaries to solve this problem. Carpellary nectaries are equally inconvenient to the plant when trends to hypogyny occur: in these cases a special pore brings the nectar onto the receptacle, as in the Zingiberaceae, Bromeliaceae and most Musaceae, which have been able to remain monoclinous. In *Musa*, the isolated male flower retains a non-functional ovary, which has become a giant nectary. The positioning of this nectar pore on the receptacle is also important, and may influence other floral structures and functions.

In the Malvales, nectar secretion is placed on the inner face of the calyx, perhaps as an *ad-hoc* regulator of visits to pollen flowers, which are ill-timed in some instances (*Tilia*). This position has apparently hindered development towards sympetaly, and thus towards the development of sophisticated tube-shapes of the corolla. Similarly in the Malpighiaceae, oil glands or nectaries in this position have required the maintenance of narrow petal claws to aid accessibility to the reward. Another example concerns the placing of nectar disks, which may influence the development of polyandry, in some cases centripetally and in others centrifugally.

SECONDARY POLLEN ATTRACTION, AND ITS CONNECTION WITH DICLINY

Specialised pollen flowers (non-anemophilous) which have arisen secondarily in higher families, the use of nectar having been dispensed with, are attractive to beetles, some syrphids and bees, but exclude pollinators which only take liquid food (Lepidoptera, birds). In these, the stamens are conspicuous, often numerous, contrasting yellow with blueish petals. Here, separate female flowers would lose attractivity, especially when pollen odour is also an attractant. It is thus not surprising that in this pollen-rich class, some genera (as *Plantago*), showed a further development to anemophily (this also probably happened formerly in the Amentiferae). Some Compositae did so, at the same time developing from the monoecy usual in this family to a form of dioecy, with male and female capitula of different construction (*Ambrosia*), which has been partly influenced by seed-dispersal requirements.

The reverse switch, from anemophily to entomophilous pollen flowers happened even in some bambusoid forest grasses, and perhaps in some bamboos. These grasses were already diclinous, but the nearness of males to females in dense stands allows entomophily in the unattractive females (also in some southern *Lithocarpus* in *Quercus*).

Both dicliny and dichogamy are unlikely to be successful in flowers with pollen attraction. However, such flowers with a potentially unattractive phase or sex have evolved other means of successful cross-pollination which allow dicliny to operate successfully. Gottsberger found cases in nectarless Brazilian Clusiaceae and Dilleniaceae which are probably relics from the archaic beetle pollination described earlier. The beetles eat pollen, but also petals, and other floral parts including the gynoecia, or lay eggs in the ovary, and thus attraction to both sexes in a diclinous flower occurs and cross-pollination results. In such parasitised *Davilla*, dicliny is effective, and it certainly exists in *Clusia* spp. in which the stigmas disseminate the same fruit-like odour as the stamens.

In other nectarless Dilleniaceae (perhaps original pollen-flowers), female flowers produce food-pollen for bees, which is functionally inviable. In *Dillenia aurea* sweet edible petals replace the function of nectar, and encourage bird-pollination. Yet another mechanism that allows attraction to the female flower in secondarily nectarless plants that are diclinous is found in *Begonia*, in which stigmas apparently mimic stamens, and thus practise successful deceit on bees. A comparable syndrome in strongly dichogamous (protandrous) flowers occurs in *Exacum affine* (Gentianaceae), in which the large poricide yellow anthers are emptied by vibration in the male phase within one week. However, they remain fresh and visually attractive in the blue flower throughout the female receptive phase, thus probably attracting bees by deceit. The same may be assumed for *Saintpaulia*, and some Commelinaceae (*Dichorisandra*).

Although we have some knowledge of the varied mechanisms by which diclinous and dichogamous flowers without nectar achieve cross-pollination, we are still ignorant of the techniques used by many, for instance dioecious Australian *Solanum* spp. and work in the Antipodes on pollen flowers in general may well prove rewarding, for this area has a strikingly high incidence of dicliny.

POLLINATION AND FERTILISATION, AND THEIR INTERACTION

The study of floral ecology usually halts once the transport of pollen to the stigma has been achieved, and at this conference one is inclined to leave aside the internal sexual processes that follow. Sometimes, however, it is necessary to proceed further. For instance, the legitimate pollinator may have to change the nature of the stigmatic surface; or the pollen grains on the stigma surface may interact (for instance through density effects, or inhibitory exudates). There is however, yet another aspect of this problem. During this paper, I have centred attention on dicliny, originating from causes other than the sheer promotion of cross-pollination. If we inspect classical mechanisms, these are assumed to promote cross-pollination and restrict self-pollination. However, these latter aims are not strictly comparable. Whereas cross-pollination concerns long-term advantages of a genetical nature (especially the promotion of heterozygosity and hence heterosis), the second also concerns short-term advantages such as the prevention of immediate damage to the stigma and ovules, and the quantity and quality of the offspring.

The avoidance of self-pollination by dichogamy, herkogamy or heteromorphy can concern all three primary barriers relating to better cross-pollination as such. They can, however, also be secondary regulators to prevent waste through unwanted 'nonsense' pollination, which is ineffective (illegitimate, wrongly timed etc). This waste may not only be of pollen, but also of stigma sites and ovules.

In the case of self-incompatibility, dichogamy does not represent a superfluous duplication, as is often said, but should be thought of in terms of 'spillage and spoilage'. Firstly, these three devices serve pollen economy by directing pollen flow to a suitable stigma (not always effective in heteromorphy). Secondly, ineffective pollen (for instance resulting from selfing) must not only be considered as a spilt, inert substance. Not only does simple occupation of the stigma surface occur, but also damage to the reactive surface of the stigma (as in some orchids). Too high a level of deposited pollen, whatever its nature, reduces seed-set in *Juglans*, although this may be due to damage by selfed pollen. In the case of *Plantago* already discussed, this aspect deserves attention in connection with the differences in monocliny and dicliny between different geographical regions.

As well as the well-known influence of growing pollen tubes on organs concerned with fertilisation and seed-formation (as for instance demonstrated by pseudogamy), we also know of direct signals to these organs, chemical and electric, derived from pollen and pollination. These are found in both higher and lower groups, and are especially pronounced in orchids, which show strong integration between functions of pollination, fertilisation and embryony. We may also expect reverse signals, for instance changes in flower colour, or wilting associated with successful embryony (*Ranunculus glacialis*). With respect to stigmatic closing and abscission as well as wilting, such changes are well-known in the orchids. In this group, rapid post-pollination changes (in otherwise long-lived flowers) assist the flower in its highly refined but chancy pollen economy and visitor economy, by avoidance of superfluous visitors.

On this subject of regulatory signals, which can also concern the timing of nectar and odour release, and is indeed central to the thesis of this paper, much more thorough work is required. Floral ecologists in the future will require this background of engineering, using techniques which at present are pursued by (for instance) Kullenberg. Perhaps it is here that one main trend in future work lies.

REFERENCES

BAKER, J. G., 1963. Evolutionary mechanisms in pollination biology. *Science, 139:* 877–883.

DARWIN, C., 1877. *The Different Forms of Flowers on Plants of the same Species.* London: Murray.

van der PIJL, L., 1972. *Principles of Dispersal in Higher Plants.* Berlin: Springer-Verlag.

SMALL, J., 1915. The pollen-presentation mechanisms in the Compositae. *Annals of Botany, 29:* 457–470.

STEBBINS, G. L., 1974. *Flowering Plants: Evolution Above the Species Level.* Harvard, Cambridge, Mass.: Belknap Press.

STELLEMAN, P., 1978. The possible role of insect visits in pollination of reputedly anemophilous plants, exemplified by *Plantago lanceolata* and syrphid flies. In A. J. Richards (Ed.), *Pollination of Flowers by Insects:* 41–46. London: Academic Press.

Evolutionary shifts from reward to deception in pollen flowers

St. VOGEL

Botanisches Institut, University of Vienna, Austria

The paper underlines some evolutionary trends in pollen flowers, supported by several new instances. Advanced pollen flowers, being strictly melittophilous, offer a pollen surplus to female bees whose activity is exclusively directed to this reward. Anthers usually advertise themselves in the open flowers by direct visual and tactile stimulation, releasing harvesting behaviour to which the cross pollination contrivances are specially adapted. True pollen abundance results from polyandry or—in oligandrous flowers—from the enlargement of the anthers (mostly poricidal, discharged by vibration). As a way to secure sufficient pollen for fertilisation, heteranthery occurs, separating cryptic functional stamens from showy feeding stamens, to which the advertising syndrome is restricted. The latter show a strong tendency to pretend pollen copiousness, or to replace fodder pollen completely by structures mimicking plenty (full anthers). A similar system prevails in isantherous pollen flowers with few and small anthers. The attention and collecting movements of visitors are distracted from true anthers to dummies on their filaments, thus preventing exhaustive pollen exploration. While such partially deceptive flowers maintain visits by a modest but real harvest, 'pollen flowers' of orchids take the risk of total deception by operating with powderlike hair tufts or empty granular pseudopollen.

KEY WORDS:—pollen flowers—feeding pollen—*Solanum*-type—heteranthery—anther dummy—pseudopollen—polyandry.

As is well known, deceptive attraction of insects by flowers is widely distributed among Angiosperms. Generally, imitations of objects other than food prevail, and deception of innate instincts is more frequent than that of behaviour governed by learning processes. Thus, bee flowers, which mimic or fallaciously promise nutrient rewards are a minority and have in part yet to be fully authenticated. This is especially true for Sprengel's 'Scheinsaftblumen', false nectaries and the like. Deceptive mechanisms no doubt also occur among pollen flowers. In the following account I want to show that this kind of deception has been unjustly neglected so far, and that several flower features incompletely understood up to now can be thus explained. Both well-known phenomena and new observations will be quoted.

Let us first define pollen flowers. In an anthecological sense, these are flowers which offer a surplus of pollen as the only reward to their insect visitors, instead of nectar or other substances. In most cases, the perianth is cup-shaped. Copious pollen is presented in a showy androecium, anthers and pollen advertising themselves optically, at least at close range, and the activities of the visitors are immediately directed to them. The flower mechanism is such that pollination is effected by this special behaviour. Consequently, nectar flowers which are additionally or passively also explored for pollen, should not be called pollen flowers;

nor should nectarless flowers visited not for their pollen, but for other reasons (gamokinetic flowers, kettle traps, wind flowers, autogamous flowers with reduced nectaries etc.).

The phenomenon of pollen flowers is varied, and as a whole insufficiently investigated. At least three basic constructions, independently realised in parallel by unrelated families, and different in age and evolutionary state, may be distinguished. Preliminarily, we may call them *Magnolian*, *Papaver* and *Solanum* types.

(1) The *Magnolian* type probably represents the most ancient pollen flowers, relics of epochs preceding nectar flowers (sometimes now possessing a meagre nectar secretion or a trap-like perigon). The copiousness of pollen production in this group is due to primary polyandry, inherited from the anemophilous pteridosperm ancestors. The pollen is mostly shed by short stamens on to the receptacle, and there directly eaten by beetles (cantharophily). This type, together with 'mess and soil' pollination (van der Pijl) prevails in the Magnoliales and Nymphaeales; the flower aggregations of many Araceae, in an analogous sense, may also be classed here.

(2) The *Papaver* type is also polyandric. The abundant pollen is not shed but is exposed in well stalked anthers. The pollen is—at least to-day—predominantly utilised by (female) bees which do not eat it, but forage for their offspring. The bees gather the pollen by swallowing (*Xylocopa*), brushing, vibrating or scrambling round the flower, thus affecting 'mess and soil' pollination. Therefore, this type must be regarded as a melittophilous specialisation. Pollen-eating beetles and flies may join in, but are of lesser importance as pollinators. Because of this haphazardous and wasteful pollination, the *Papaver* type also has somewhat primitive traits. It certainly arose heterogeneously, as can be deduced from differences in the androecial ontogeny, and the secondary origin of some of these polyandrous androecia. Excepting truly multistaminate members, such as Papaveraceae and Ranunculaceae (*Ranunculus asiaticus*, *Anemone*, *Thalictrum*), the polyandry of rather primitive families belonging here (Dilleniaceae, Actinidiaceae, Cochlospermaceae, Ochnaceae, Paeoniaceae, Cistaceae, Guttiferae, Tiliaceae Begoniaceae), mostly having pollen flowers, may be derived from a whorl of a few multisporangiate sporophylls from cantharophilous times. Others like *Rosa*, *Rubus* sect. *Anoplobatus*, pollen-flowered Aizoaceae and Portulacaceae (*Glottiphyllum*, *Portulaca*), Mimosaceae (*Acacia* spp., *Mimosa*) and Cactaceae (*Opuntia*), members of mainly nectariferous families, appear to be secondary, adaptive transformations of earlier polyandrous or oligandrous nectar flowers. Sometimes, the boundary is not sharp and there are still nectarogeneous transitions to pollen flowers as in *Cistus*, *Paeonia*, Theaceae, Guttiferae, many Aizoaceae and Cactaceae, where the visiting bees may show both sucking and pollen foraging behaviour.

(3) The *Solanum* type, finally, comprises all oligandrous pollen flowers. In this 'modern', strictly melittophilous, often zygomorphic specialisation, the visitor clasps the androecium in a fixed position. As a rule, the few anthers are enlarged, thus becoming more showy and capable of producing excess pollen; they are mostly sessile (by neotenic shortness of filaments) and poricidal. The powdery pollen is released in small portions as a 'cloud' by vibrating movements of the visitors, thus dusting their bodies, and is subsequently indirectly harvested by combing. Because of this overall dusting of the visitor, the position of the small

stigma is independent from that of the anthers. Although some of these oligandrous pollen flowers, in the Dilleniaceae (*Schumacheria, Hibbertia*), Ochnaceae (*Luxemburgia, Ouratea*), Primulales (*Ardisia, Lysimachia, Cyclamen, Dodecatheon*), Melastomataceae, Malpighiaceae and Commelinales, probably never possessed a nectary and changed directly from 'dry', polymerous ancestors, many representatives of the *Solanum* type are derived from oligandrous nectar flowers, sometimes still showing a rudimentary, non-functional nectary. Actinorphic pollen flowers of the *Solanum* type occur, besides *Solanum*, in *Ramonda, Sabatia, Dichorisandra, Hypoxis, Curculigo, Barbacenia, Galanthus, Leucojum, Calectasia, Sowerbia, Luzuriaga, Aristea* and the Cyanastraceae, Mayacaceae, Rapateaceae. Zygomorphic pollen flowers occur in *Saintpaulia, Alonsoa, Chironia, Orphium, Exacum, Henriettea, Pyrola*, the Byblidaceae and the Roridulaceae. All these have a floral appearance in common, which is characteristic, but the significance of which was hitherto not fully recognised. The papilionaceous pollen flowers (many Genisteae) and some louseworts (*Pedicularis*) with a concealed pollen load are an aberrant variety of this ecological class, also exclusively adapted to modern Apidae.

Contrary to nectar flowers, where the evolutionary trend towards oligandry was a positive progression, reducing pollen waste as well as leading to a more precise pollination, oligandry counteracts the biological principle of pollen flowers, because pollen quantity decreases. Only a few flexible genera were able to return to polyandry on becoming secondary pollen flowers (*Swartzia, Cleome hirta*), the bulk remaining oligandrous or, as in the Commelinaceae, even continuing staminal reduction.

As we have seen, in most cases the anther size increased in a compensatory way, thus creating sufficient additional pollen, and more visibility (*Tulipa*). Other oligandrous taxa on becoming pollen flowers were incapable of effecting that increase. But almost no oligandrous pollen flower exists which has small or unspecialised anthers in the mode of nectar flowers; nectarless flowers of this appearance, like *Anagallis arvensis* or *Lysimachia nemorum*, are autogamous. To maintain cross pollination, several methods have been employed by such flowers to escape from plainness, and from losing too much pollen for insect food, for instance accumulation, as in the pollen flower capitula of *Plantago, Sambucus, Aruncus* and *Filipendula*. Two other methods, heteranthery and deceit and especially the combination of both, are elaborate.

In isantherous flowers (with monomorphic anthers), anther and pollen dummies, formed by different parts of the stamen or surrounding regions, mimic pollen copiousness by showing a yellow and swollen appearance. These anther substitutes not only have the function of increasing attractivity, but may also distract the visitor's approach and foraging activity from the true anthers or thecae, thus protecting the latter from exhaustive exploration. In *Xyris, Bulbine, Arthropodium, Stypandra, Narthecium, Tradescantia* and *Anagallis* species, false anther substitutes are made of hair tufts, mostly on the filament; swollen yellow filaments give the small anthers of *Dianella*, and some *Cleome species, Keraudrenia* and *Conandron* a more promising make-up, whereas in many Melastomataceae (*Loreya, Blakea* and *Axinaea*) and Commelinaceae (*Zebrina, Setcresia*), the connective is the mimetic part. *Meriania* (Melastomataceae) is certainly one of the most curious examples. In *M. longifolia* the conspicuous yellow parts on the lower end of the one-sided androecium are not what they are taken for by a superficial observer, nor by the Carpenter bees working these flowers, namely anthers full of pollen. In fact, they are dummies formed by connective appendages, the true poricidal thecae being almost concealed below the other dull-coloured stamen parts. *Xylocopa*, while vibrating the dummies in a downward-faced position by clasping them, unintentionally causes a pollen cloud to leave the anther pores and dust their abdomens.

We should add that in the large-anthered flowers mentioned above, deceit is also playing a role. Most poricidal anthers, provided with persistent, papery, bright yellow pollen sac walls (often with a reduced endothecium), keep their swollen form from the time before dehiscence (in case of protogyny) until long after their complete evacuation, thus provoking visits and ensuring cross pollination for a much longer time than pollen is available. "Deceptive stages" may thus be added to the stage of genuine reward. The interesting case of *Begonia* may be mentioned in this connection, although it approximately belongs to the *Papaver* type. The staminate flowers produce pollen as a true reward. The pistillate flowers, on the other hand, have nothing to offer, but are visited as well. Their means of attraction must be deceptive: the voluminous yellow stylodia mimic the androecium of the male flowers and thus release gathering movements which, of course, effect pollination.

The well known heteranthery is another means of solving the problem of pollen flowers with little pollen, i.e. preserving sufficient pollen for fertilisation, especially in oligandrous flowers. Contrary to heterostyly, it is strictly confined to pollen flowers, although in some cases (*Lythrum*, *Aneilema*, Pontederiaceae) there are affinities between both kinds of staminal dimorphism. Even today, the occurrence of two different sets of stamina is explained as a mere consequence of zygomorphism. Nevertheless, it is also functional, a division of labour which separates the showy feeding anthers from fertilising anthers (reward from pollination). The two destinations of pollen are here unmistakably demonstrated, as in most cases the advertisement syndrome is confined to the 'fodder' androecium, provoking the visitor's exclusive approach, while the pollination stamina or anthers (and sometimes even their pollen) are inconspicuous or of the same colour as the perianth. They are exposed so as to be touched unconsciously by insect body parts less accessible to grooming, often in accordance to the stigma position.

Although this kind of pollen economy prevails in combination with oligandry, it is not completely absent in polyandrous pollen flowers. The rare cases of actinomorphic heteranthery are found in *Mollea* and *Lagerstroemia*. *Mollea speciosa*, a tiliaceous tree of the Amazon, has around a central bundle of showy feeding anthers which bear yellow pollen, five fascicles of pollination stamina with brown anthers of withered appearance and green pollen. In *Lagerstroemia indica* (Lythraceae) the same principle occurs. The six long, episepalous fertilising stamens as well as the style are purple like the petals, curved so that the bee's back is touched.

Among zygomorphic polyandrous flowers, *Amourexia* (Cochlospermaceae), *Hibbertia* sect. *Hemipleurandra* (Dilleniaceae), *Swartzia* (Caesalpiniaceae) and *Cleome hirta* (Capparidaceae) are heterantherous. All have a large bouquet of feeding stamina, but relatively few fertilising stamina. In *Swartzia*, some species—

as in the related genus *Aldina*—are isantheric, regarded by Cowan in his mono-graph (1968) as the most derived, through loss of the few big fertilising stamens present in the rest of the genus (in my opinion, the reverse is more probable, i.e. that it was the first stage, from which fertilising stamens originated; these, once in function, could hardly have disappeared again). While *Swartzia* has no colour differences, in *Cleome* fertilising stamina, anthers, and pollen are violet like the perianth, and the feeding androecium is bright yellow, visually assisted by a basal field of the same colour, on the two upper petals.

In the voluminous feeding androecium of *Swartzia*, the anthers are small and poor in pollen. Nevertheless they are the only ones explored; euglossid and *Centris* females work them by vibration, disregarding the rich lower anthers. In this case we see a feeble beginning of a trend from reward to deceit among heter-antherous flowers. The partial or total replacement of true anthers and pollen by dummies is similar to those of isantherous flowers but, of course, confined to the fodder stamina. Returning to Tiliaceae, the nodding pollen flowers of *Sparmannia africana*, well known for their haptonasty, may also be regarded as heterantherous. The showy peripheral staminodia shamming pollen copiousness by yellow fila-ment knobs, play the role of feeding anthers; visitors which try to explore them touch the fertile androecium and stigma with the ventral side of the abdomen. Still undescribed cases of heteranthery in zygomorphic polyandrous flowers with a strong tendency to fallacious pollen presentation are found in the Lecythidaceae. In *Gustavia augusta*, the totally fertile androecium shows only a slight separation of the two sets of stamina. In *Couroupita*, strong differentiation takes place between a brush-like dorsal fertilising area around the short style, and a lower set consisting of a lip-like androphore covered on its whole inner face by big yellow, polliniferous nutrition stamina. The visitor working them does not pay attention to the dorsal set, but touches it with its back. Final steps are reached in *Berthol-letia*, *Lecythis*, and *Eschweilera*, where the androphore merely bears staminodial warts and crests. In *Bertholletia* these are yellow and exclusively (but of course vainly) brushed by the deceived visitor.

In oligandrous pollen flowers, heteranthery also starts in a mutualistic manner, in which the feeding anthers offer true but often sterile fodder pollen, and proceeds to partial deception. In the Melastomataceae, *Cassia*, *Solanum* and other cases, gradual emergence of anther dimorphism can be observed on comparison of different species. In the Melastomataceae the basic construction leading to heter-anthery is an isostemonic fascicle bent upward opposite to the style, as in *Topobea*, *Osbeckia* and *Rhexia*. In most species with anther dimorphism (*Melastoma*, *Tibouchina*, *Dissotis*), five episepalous stamina maintain the fertilising function, curving downwards, parallel to the style, and have a cryptic colour; while five episepalous stamina form an upper fascicle bearing smaller anthers but showy connective dummies (as are also found in isantherous species: *Meriania*). Thus the differentiation of the two sets is not effected by structural zygomorphism. Their nastic separation in an upper and lower fascicle is caused by differences in geotro-pism. Within the genus *Cassia*, once again a poricidal vibration mechanism, the gradual transformation of an isantheric androecium to heteranthery, governed by autonomous zygomorphism, is well known. In *Solanum*, a mainly isantheric genus, heteranthery probably arose more than once. The sections *Nycterium* and

Lycianthes contain species (*S. sisymbriifolium*) with one of the five anthers only slightly enlarged; in *S. rostratum*, *S. vespertilio* and *S. linii*, the same anther has become a big, bellows-like fertilising organ, with an alternative position to the stigma as in some *Cassias* (enantiostyly). In the similar *S. citrullifolium*, the big anther has the same purple hue as the perianth, while the rewarding anthers remain yellow. In the monocotyledonous, *Cyanella lutea* (Haemodoraceae), this principle is exactly repeated. Other species, such as *C. alba*, have isantherous pollen flowers. Parallels have been found in South African *Anthericum* and *Chlorophytum* species and also *Heteranthera* (Pontederiaceae). Part of Old World Malpighiaceae have isantheric pollen flowers (*Acridocarpus*); other ones have developed one large median, exserted, fertilising stamen parallel to the style (*Hiptage*). In the partly isantherous American genera *Malpighia*, *Banisteria*, *Stigmatophyllum* and several others, heteranthery arose independently; a pair of opposite lateral stamens, again parallel to two styles, is exserted for fertilisation. In these groups, bearing oil flowers, the attractive oil glands (elaiophores) on the sepals may be regarded as an additional means of preventing the precocious and exhaustive emptying of the anthers by the visitors (similar to the case of— isantherous—*Calceolaria* and nectarless species of *Lysimachia*). Heteranthery can also be assumed in the diclinous flowers of *Anacardium pumilum*, where the staminate flowers have but one large fertilising stamen in addition to a set of short, probably feeding stamina, which are also present in the pistillate flowers, thus insuring cross pollination by providing them with (sterile) pollen reward.

In most of the preceding instances, heteranthery is mutualistic. Pollen is yielded as a true reward. A shift to deception can be observed especially in oligandrous flowers with sticky pollen in small anthers. In these, imitations strengthen the attractivity of the fodder stamina in a similar manner to those in deceptive isantherous flowers, but the anthers may completely disappear. In *Verbascum*, also a fine example of the gradual emergence of androecial dimorphism in one genus, the small upper stamina of the heterantherous species such as *V. thapsus* or *V. thapsiforme*, appear—at least to bumblebees—to have plenty of pollen, mimicked by the woolly filament hairs. These—as in *Tradescantia*—erroneously thought to be food by earlier botanists ('Safthaare'), are scratched and combed by the visitors in the belief that they are harvesting pollen, and are often to be found in the corbiculae together with real pollen grains. In Commelinaceae, a family devoid of nectaries and probably consisting throughout of pollen flowers, similar cases of partial deception occur in the heterantherous flowers. *Tripogandra grandiflora*, with regular white flowers, has normal whitish fertilising anthers on short filaments similar in length to the style, and long hairy feeding stamens with broad yellow connectives. In some zygomorphous genera, the fertilising stamina are the longer ones and sternotribic.

In *Commelina coelestis*, the three luxuriant advertising connectives, of a striking yellow in contrast to the blue perigon, are the centre of attraction. They are not eaten, as claimed in the literature, but yield a minute quantity of pollen in their thecae. The visitor, while foraging, touches the totally blue, non-contrasting fertilising stamina. These kinds of deceit may be derived from isantheric *Zebrina* and other actinomorphic genera where all stamina have moderately enlarged connectives. In *Tinantia fugax*, instead of connective dummies, the filaments of the advertising stamina are provided with tufts of yellow moniliforme hairs, once again mimetics of pollen abundance, the adjoining anthers themselves being very poor pollen producers. A final step in this development is shown by the enantio-stylic flowers of *Cochliostema*, the biology of which has long been an enigma. Nutritional anthers and pollen have completely disappeared and are replaced by a yellow hair tuft. Fertilising pollen is produced by the twisted thecae of two fertile stamina and hidden in a bellows-like perigon-coloured pseudo-anther (a formation of filament wings), probably shed by vibration, when the visitor tries to take the pollen from the dummy.

As already pointed out, in all mentioned cases of deceptive pollen flowers, deception is only partial, because a small quantity of pollen as true reward will usually be gathered, either from the monomorphic anthers in the isostemonic flowers, or true fodder pollen in the heterantherous flowers. Even in *Cochliostema*, visits are moderately rewarded by some of the fertilising pollen obtained by the insect afterwards when grooming its body. To maintain at least a certain degree of reward seems necessary, as the intelligent hymenopteran guests would otherwise soon cease further visits, and insufficient cross pollination would result.

An exception apparently occurs in the Orchids. As in other modes of deceit, this family also makes a total gamble in this mode of flower parasitism. Of course, no true granular pollen is offered, the pollinia being unsuitable for foraging. However, several nectarless genera are well known to represent a kind of pollen flower through pollen imitation. In *Pogonia*, *Arethusa*, *Calopogon*, *Oncidium pulvinatum*, *Eria* sp. and *Thelymitra crinita*, yellow hairs, mostly on the labellum, appear to function as pollen dummies. In the traps of many *Maxillaria* and *Polystachya* species these trichomes are moniliforme and 'sporulate' granular pseudopollen. In *Maxillaria* this is said to contain starch and thus to have nutritional value. This case, however, should be reinvestigated. The pseudopollen of *Polystachya pobequini*, at least, is empty and inedible. The question as to why orchids are able to risk complete frustration of their pollinators, remains open. Possibly these flowers, always a minority, derive profit from similar but really rewarding pollen flowers of other more common taxa in the same biotope.

In summary, through the comparison of these extreme cases with archaic cantharophily, the whole range of pollen flower evolution becomes evident. In early times, still represented by the Magnolian type, pollen flowers seem to have been a direct consequence of primitive polyandry. The nectarless polyandric flowers of the dillenioids, rosoids etc., also an ancient development derived from cantharophily in part, underwent a coevolution with pollen-utilizing Apidae leading to the recent but still primitive melittophilous *Papaver* type. The oligandrous flower of the *Solanum* type, being exclusively adapted to bees, is a late-comer, presupposing an established bee fauna with an elaborate visiting behaviour. By utilization of these instincts, heteranthery and other pollen-saving deception practices evolved, the total deceit of orchid 'pollen flowers' being a final stage in the development of these tendencies.

REFERENCES

COWAN, R. S., 1968. *Flora Neotropica, Monograph No. 1: Swartzia*. New York & London: Hafner.
DAUMANN, E., 1971. Zum Problem der Täuschblumen. *Preslia, 43:* 304–317.

MACIOR, L. W., 1964. An experimental study of the floral ecology of *Dodecatheon meadia*. *American Journal of Botany, 51:* 96–108.

MOORE, L. W., 1960. *Tripogandra grandiflora* (Commelinac.). *Baileya, 8:* 77–83.

van der PIJL, L. & DODSON, C., 1966. *Orchid Flowers, Their Pollination and Evolution.* Coral Gables: University of Miami Press.

THIEN, L. B. & MARCKS, B. G., 1972. The floral biology of *Arethusa bulbosa, Calopogon tuberosus* and *Pogonia ophioglossoides*, Orchidaceae. *Canadian Journal of Botany, 50:* 2319–2325.

VOGEL, St.,1974. Ölblumen und ölsammelnde Bienen. In E. Reihe (Ed.) *Tropische und Subtropische Pflanzenwelt*, Vol. 7, Wiesbaden: Franz Steiner.

VOGEL, St., 1975. Mutualismus und Parasitismus in der Nutzung von Pollenträgern. *Verhandlungen der Deutschen zoologischen Gesellschaft, 1975:* 102–110.

WERTH, E., 1956. *Bau und Leben der Blumen.* Stuttgart: Enke.

ZIEGLER, A., 1925. Beiträge zur Kenntnis des Androeceums und der Samenentwicklung einiger Melastomataceae. *Botanisches Archiv. 9:* 398–467.

The physiology of some sapromyophilous flowers

B. J. D. MEEUSE

Botany Department, University of Washington, Seattle, Washington, U.S.A.

An effort will be made to provide some physiological and biochemical background to St. Vogel's beautiful descriptions of the pollination-syndrome in certain arum lilies, and to L. van der Pijl's pioneer experiments on the role of light and darkness in the opening of some flower-species. The highly thermogenic, cyanide-insensitive respiration which characterises the so-called appendix of *Sauromatum* on the first day of flowering is triggered by a hormone ('calorigen') which originates in the primordia of the staminate flowers and begins to leave these about one day before opening-time. The formation (or the release?) of the hormone is, in its turn, controlled by the particular light/dark regime to which the developing inflorescence has been exposed. A single dark-period of at least 6 hours' duration, given at the right moment to inflorescences first allowed to develop in constant light, leads to the characteristic 'respiratory explosion' about 45 hours later. It will be shown that the opening of certain flowers, often thought of as a very simple act, may likewise depend on a series of 'programmed' events set in motion two days before actual opening-time.

KEY WORDS:—aroid—*Sauromatum*—calorigen—respiratory explosion—dark period—sapromyophilous.

CONTENTS

INTRODUCTION

It is probably fair to say that aroids began to attract the attention of scientists 2000 years ago. The drawing of *Arum maculatum* or *Arum italicum* found in the A.D. 412 version of the famous herbal of Dioscorides (*De Materia Medica*) may well be based on sketches by Crateuas, going back to 120 B.C. (Prime, 1960). Studies on the thermogenicity of aroids are about 200 years old, for it was in 1778 that the great Jean Baptiste de Monet, Chevalier de Lamarck, recorded the spectacular temperature-rise displayed by the inflorescences of *Arum maculatum*. Floral biology, which started a little later (Sprengel, 1793) has given us an understanding of the survival-value of the heat-development (Knoll, 1926; Vogel, 1963; see below).

It is both intriguing and depressing that an area of study with such impressive historical credentials—potentially a happy meeting-ground for floral ecologists on the one hand and plant physiologists and biochemists on the other—remains rife with controversy. Undeniably, linguistic barriers have played a role in creating this stage of affairs. A sad testimonial is provided by Knoll's splendid contributions on *Arum*, published in 1926, which remained virtually unknown in English-speaking countries, until, in 1960, Dormer drew attention to them in "The truth about pollination in *Arum*". Although linguistic barriers are now less substantial, new ones have sprung up as a result of the fact that biochemically oriented investigators no longer speak the same scientific language as colleagues of a "natural history" bent. It is the purpose of this article to demonstrate that the two groups can provide *each other* with valuable clues for further research and understanding. The approach taken here will be comparative as well as analytical and predictive; a springboard will be provided by the situation in *Arum*, which is not only of historical importance, but happens to highlight several problems as well.

POLLINATION EVENTS IN THE GENUS *ARUM*

Knoll (1926) has correctly interpreted the inflorescences of *Arum nigrum* and *Arum maculatum* as 'trap-flowers' which 'collect' their pollinators (small beetles and midges, respectively) in a floral chamber into which the visitors fall when they are denied a foothold on the slippery surface of the spathe, and of the appendix: the heat and smell producing osmophore, *sensu* Vogel, which represents the naked, sterile upper end of the fleshy central spadix. The stiff hairs or barrier-organs at the entrance of the floral chamber serve as a sieve or collander, keeping out larger animals such as bluebottle and greenbottle flies which might also be attracted by the smell of the appendix. The inflorescences display a very pronounced pro-togyny; the pistillate or 'female' flowers surrounding the base of the fleshy spadix in the floral chamber are receptive when the appendix is active in producing its heat and smell. The visitors, some of which may be carrying pollen from another *Arum*-inflorescence which was in the pollen-shedding or 'male' stage, feed on a stigmatic secretion produced by the pistillate flowers and pollinate them in the process. It is only several hours later that the staminate or 'male' flowers, bunched together in a ring higher up on the spadix, dehisce and shower the captive insects with a rain of pollen, so that the next day these animals are covered with pollen. At that time, the barrier-organs have wilted so that they are no longer functional, while the stigmatic secretion as well as the heat and the smell have disappeared; the visitors leave, with the pollen which they have received and may visit another inflorescence which is still in the smelly, attractive, 'female' stage. It would be difficult to find a better-designed, more foolproof mechanism to insure cross-pollination than the one just described. For the plant physiologist, the interest lies in the importance of timing-mechanisms.

FUNCTION AND TIMING OF THE HEAT-DEVELOPMENT IN *ARUM* LILIES, AND THE RELATIONSHIP BETWEEN HEAT AND SMELL

Knoll convincingly demonstrated that the heat *per se* does not act as an attrac-tant: models of *Arum*-inflorescences with small, heat-providing light-bulbs in the floral chamber failed to attract pollinators, whereas models provided with a mixture of rotting blood and glycerol (an antidesiccant) did. The biological function of the heat is to aid the evaporation of odoriferous compounds (a mixture of various amines, ammonia, and indole or skatole). The heat-production, which in *Arum maculatum* can lead to a temperature-difference between the appendix surface and the environment of about 15°C (in other aroids it may be even higher: 22°C in *Alocasia pubera* and *Schizocasia portei*), is based on a starch-consuming respiration process unparalleled in the plant kingdom, so that in the course of a single day the dry weight of an appendix may fall from about 32% to about 6%. When this phenomenon reaches its peak, the intensity of the metabolic process is comparable to that of a flying hummingbird; oxygen-consumptions as high as 72,000 mm³ of oxygen per gramme wet weight per hour have been recorded (Lance, 1972). As explained below, respiration in this case is of a special 'uncoupled' and cyanide-insensitive type so that very little ATP or other high-energy compounds are generated; practically all the energy locked up in the respiratory substrate (starch) is expended in the form of heat, an unusual situation which in this case, however, has a high survival-value. I have demonstrated the heat-production visually by putting on the appendix a thin film of a vaseline-like mixture of liquid crystals; since the manner in which such crystals 'handle' light is temperature-dependent, a striking sequence of colours ranging from copper-red to peacock-blue manifested itself as the appendix, in the process of heating-up, passed through the 25°–28·5°C range. Smith & Meeuse (1966) as well as Chen & Meeuse (1971) have analysed (to some extent) the mixture of odoriferous compounds emanated by arum lily inflorescences of various species. A high degree of specificity is evident; thus, *Sauromatum guttatum* produces, among other things, large amounts of indole, while *Arum dioscoridis* yields skatole instead. The differences are usually quite obvious to the human nose; e.g. inflorescences of *Arum orientale* have the characteristic smell of spent fire-crackers (due to the presence of sulphur compounds?) while those of *Arisarum vulgare* are reminiscent of dead freshwater fish. In consonance with the different chemical composition of the emanations, the visitor-spectra too are different; thus, *Helicodiceros muscivorus* is pollinated almost exclusively by large flies; *Arisarum proboscideum* (which displays beautiful mushroom-mimicry: Vogel, 1973) by fungus-gnats; *Dracunculus vulgaris* by beetles. It is plausible that there is also a good correlation between the life-style of the pollinators and the hour of the day at which the production of heat and smell reaches its peak; in *Arum maculatum*, that peak occurs in late-afternoon or early-evening, in *Sauromatum* around noon. Obviously, sympatric species of aroids avoid competition for pollinators.

Leick (1915) was probably the first investigator to compare various species of Araceae with regard to the number of heating-periods through which each inflorescence goes during the flowering-sequence. It is here that the enormous advantage, in terms of energy, of the 'invention' of the floral trap becomes obvious. Certain primitive Araceae lack a true, functional floral chamber and possess a spadix which is covered with small hermaphroditic flowers from top to bottom. Protogyny is still very much in evidence, and the production of heat and smell must therefore occur at least twice: the first time to attract pollen recipients. In contrast, an inflorescence with an effective trap, such as that of *Arum maculatum*, essentially secures for itself a 'captive audience' and can therefore restrict itself to only one pronounced heating-up period at the time the pistillate flowers are receptive, although a small second heating-peak will still manifest itself when the pollen reaches maturity. A myth which has persisted for some time among certain floral ecologists is that the expenditure of much heat is necessary for the production of amines and other chemical attractants. It is based on a complete misunderstanding of thermodynamics. In plant cells, 'heat' is essentially a useless form of energy, a waste-product, and there are mechanisms in both plant and animal cells for trapping the energy of the respiratory substrate as ATP rather than permitting early heat-production. The appendices of arum lilies are truly exceptional. It can be demonstrated that the bulk of their heat-production occurs in the decomposition (into water and oxygen) of hydrogen peroxide which arises as the final product of their peculiar respiration-process. It is difficult to see this as something connected with a synthetic act. It is true that a large production of heat is indicative of a large breakdown of starch, and it is a matter of simple observation that this breakdown is accompanied by the production of certain odoriferous compounds. In a quantitative sense, however, there is a large discrepancy between the two processes. The breakdown of starch can be expressed in grams while the concomitant production of amines etc. is expressed in milligrams. If the only biological 'purpose' of the starch breakdown were the generation of energy for amine production, the process would be very inefficient indeed, and it is unlikely that such a process would have been maintained in the highly selective process of evolution.

In the aroid inflorescences heat should be seen as a volatiliser. However, there are exceptional cases, such as that of *Symplocarpus foetidus* (eastern skunk cabbage) where the developing inflorescences will push themselves up through the snow even when the ambient temperature is below freezing. Knutson (1972, 1974) has demonstrated the existence of a regulatory mechanism which maintains the temperature-constancy of the floral parts, regardless of the prevailing environmental temperature: the lower the latter, the higher the intensity of respiration. Self-pollination seems to be normal in *Symplocarpus* in Iowa, although later in the season simulids and even honeybees are visitors.

The question as to whether the production of odoriferous compounds and of heat are just two aspects of one and the same process, respiration, has also received attention. An affirmative answer would imply that heat and smell *must* manifest themselves at the same time (if both are present). This has occasionally been challenged. It has also been emphasised, in some quarters, that certain Araceae produce smell, without giving off appreciable heat. The power of the latter argument is somewhat less than impressive as long as it is not accompanied by the presentation of very accurate and sophisticated temperature-measurements. Certainly, it should not be used as evidence against the idea that heat acts as a volatiliser; the tacit assumption is that every pollination mechanism has to be perfect (or 'complete'), and this is incorrect. Much stronger evidence for the functional independence of heat-production and smell would be the observation that heat can occur in aroids in the absence of smell. To the best of our knowledge, however, such an observation has never been made. If it had been, it would of course have provided very strong evidence indeed that the heat *per se* can act as an insect-attractant (see above). The production of amines by aroid appendices can indeed be seen as an aspect of respiration in the broad sense, whether it be due to decarboxylation of amino acids as suggested by Simon (1962) and Richardson

(1966), or to transamination of aldehydes, as championed by Hartmann, Dönges & Steiner (1972a, b). Free ammonia can potentially be formed in the oxidation of glutamic acid by glutamic acid dehydrogenase, an enzyme which in the appendix of *Sauromatum* reaches its highest activity when there is maximal heat-production (Meeuse, unpubl.). Further evidence for the *very* close metabolic connection between heat-production and smell (or at least between heat-production and one smell-compound, indole) has been provided by Chen & Meeuse (1972) and by McIntosh (unpubl.).

THE BIOCHEMICAL BASIS FOR THERMOGENICITY IN *ARUM* LILIES, AND THE TRIGGERING OF THE METABOLIC EXPLOSION IN THE APPENDIX

The cyanide-insensitivity of the appendix-respiration, reported above, resides in the mitochondria. When isolated from the tissue, these contain a dual pathway for respiratory electron transfer: the classical, cyanide-sensitive electron transport system which is coupled to the generation of high-energy phosphate (ATP), and a cyanide-insensitive pathway which branches off from the classical one at the ubiquinone link and is phosphorylative to a much lesser extent. When cyanide is added to respiring mitochondria in a laboratory experiment, the electrons coming from the respiratory substrate are *forced* to go through the alternate pathway only; very little ATP is generated, and the energy of the respiratory substrate quickly appears as heat. Hess & Meeuse (1967) have provided evidence that on the first day of flowering in *Sauromatum* the alternate pathway is followed even in the absence of cyanide, i.e., spontaneously. Presumably, the same situation obtains in related species such as *Arum*. It is as if a switch has been thrown, releasing heat. How is this achieved? On the basis of the pioneer work done by van Herk (1937a, b, c) with *Sauromatum*, it is now possible to postulate that the agent responsible is a plant hormone, calorigen, which is produced in the primordia (buds) of the staminate flowers and begins to leave these to move into the appendix about 22 hours before the metabolic peak is reached. Meeuse and coworkers have succeeded in concentrating and purifying two compounds with calorigen-activity. They are low-molecular, aromatic compounds of considerable stability; complete elucidation of their chemical structure and properties seems only a matter of time (Buggeln & Meeuse, 1971; Chen & Meeuse, 1975). In the purification, a bio-assay based on the formation of indole by pieces of immature appendix treated with calorigen-containing extracts has been very helpful. The long lag-time between the exposure to calorigen and the appearance of the metabolic peak argues in favour of the idea that calorigen-triggered synthesis of new enzymatic protein (as the result of an unblocking of certain genes?) is essential. This hormone would mediate the switch to the cyanide-insensitive heat-producing pathway at the ubiquinone link. This idea is now being tested by treating immature pieces of *Sauromatum*-appendix with calorigen and then adding inhibitors such as cycloheximide and puromycin which interfere with protein synthesis; these inhibitors can be added simultaneously with the calorigen or a specified number of hours afterwards (McIntosh & Meeuse, unpubl.).

Earlier, Simon & Chapman (1961) had already obtained evidence relevant to the question of protein synthesis. With the aid of electron micrographs, these authors compared mitochondria from *Arum maculatum* appendices collected on the day of heat production ('D' day), with mitochondria from earlier stages. They found the number of cristae per mitochondrion to be significantly higher for D-day mitochondria, as compared with younger ones; and, since cristae represent 'enzymatically active surface area', the results clearly favour the idea of protein synthesis during development. However, a considerable proportion of this takes place before there is any triggering by calorigen (which is also found in *Arum*; see Buggeln & Meeuse, 1971). It is also true that in *Sauromatum* there is a positive response to the addition of calorigen even in pieces of appendix so young (D minus 3 days, for instance) that the cristae are still highly underdeveloped and the amount of enzymatically active protein is low. Obviously, the metabolic explosion can be brought about even with relatively incomplete metabolic machinery, and the question of control remains open; perhaps only certain key cofactors have to be synthesised, or perhaps crucial cofactors have to be juxtaposed to other metabolites in a different manner than before, a process in which membrane-permeability changes may be very important. It is also important to keep in mind that an almost explosive mitochondrial respiration has to be 'supported' by a spectacular boost in glycolysis. Which factors play a key role in that process is a problem that has been addressed by Hess & Meeuse (1968) and by Johnson & Meeuse (1972). There is evidence that phosphofructokinase plays a major role in this boost.

THE ROLE OF THE LIGHT/DARK REGIME IN THE FLOWERING OF *SAUROMATUM* AND *ARUM*

Although van Herk's publications create the impression that the flowering-sequence in *Sauromatum* proceeds normally even in constant light, B. J. D. Meeuse and coworkers have clearly shown that this is not so. Confirming earlier observations by Schmucker (1925), they found that a regime of constant darkness prevents the development of heat and smell. Constant light permits 'half-hearted' anthesis, but the fact that this is not synchronised in a group of individuals, in contrast to the natural situation in which *Sauromatum* goes through a peak at noon as described in this article, shows clearly that an alternation of light and dark periods is required. Buggeln, Meeuse & Klima (1971) raised *Sauromatum* inflorescences in constant light until they were sure that D-day would normally have passed. Imposition of a 6-hour 'dark shot' at that time led to anthesis, with the metabolic peak occurring 42–45 hours after the beginning of the dark period. Surprised by the length of the lag-period, Meeuse then initiated a literature-search which showed that the *Sauromatum*-situation can by no means be considered unique. It is tempting to see the actual opening of flowers in general as a very simple act, controlled by such things as turgor changes in certain cells of the petals or a temperature-induced difference in growth-rate between the upper and the lower layers of a petal or tepal, as is the case in crocuses. In a number of cases, however, it is obvious that about two days before flower-opening, the bud is already 'committed' to that act. Once the 'triggering' has taken place, it is of minor importance what happens during the long lag-period. Among the best pieces of evidence is the elegant investigation carried out by Arnold (1959) on species of *Oenothera*. Normally, the flower buds open early in the evening, e.g. at 18.00 hours. When day and night are reversed, the buds continue to open at their accustomed time for two days; it is only on the third day that the new time-schedule is adopted, so that flower-buds open at 06.00 hours. In cases like this, it is again tempting to postulate the synthesis of new enzymatic protein. One could formulate the hypothesis that for the ultimate act of flower-opening, certain osmotic phenomena (turgor changes) are essential, but these may require free sugars, which first have to be produced by the hydrolysis of starch present in the buds, a process which in turn may require the presence of certain enzymes, made available by the activation of certain genes. A whole new field of investigation thus awaits investigation, and it is gratifying that studies on aroids have contributed towards that development.

REFERENCES

ARNOLD, C.-G., 1959. Die Blütenöffnung bei *Oenothera* im Abhängigkeit vom Licht-Dunkel-Rhythmus. *Planta, 53:* 198–211.

BUGGELN, R. G. & MEEUSE, B. J. D., 1971. Hormonal control of the respiratory climacteric in *Sauromatum guttatum* (Araceae). *Canadian Journal of Botany, 49:* 1373–1377.

BUGGELN, R. G., MEEUSE, B. J. D. & KLIMA, J. R., 1971. Control of blooming in *Sauromatum guttatum* Schott by darkness. *Canadian Journal of Botany, 49:* 1025–1031.

CHEN, J. & MEEUSE, B. J. D., 1971. Production of free indole by some arum lilies. *Acta botanica neerlandica, 20:* 627–635.

CHEN, J. & MEEUSE, B. J. D., 1975. Purification and partial characterization of two biologically active compounds from the inflorescence of *Sauromatum guttatum* Schott (Araceae). *Plant Cell Physiology, 16:* 1–11.

DORMER, K. J., 1960. The truth about pollination in *Arum*. *New Phytologist, 59:* 298–301.

HARTMANN, T., DÖNGES, D. & STEINER, M., 1972a. Biosynthese aliphatischer Mono-amine in *Mercurialis perennis* durch Aminosäure-Aldehyd-Transaminierung. *Zeitschrift für Pflanzenphysiologie, 67:* 404–417.

HARTMANN, T., ILERT, H.-I. & STEINER, M., 1972b. Aldehydaminierung, der bevorzugte Biosyntheseweg für primäre, aliphatische Mono-amine in Blütenpflanzen. *Zeitschrift für Pflanzenphysiologie, 68:* 11–18.

HERK, A. W. H. Van, 1937a. Die chemischen Vorgänge im *Sauromatum*-Kolben. *Recueil des travaux botaniques néerlandais, 34:* 69–156.

HERK, A. W. H. Van, 1937b. Die chemischen Vorgänge im *Sauromatum*-Kolben. II. *Proceedings K. Nederlandse akademie van wetenschappen, 40:* 607–614.

HERK, A. W. H. Van., 1937c. Die chemischen Vorgänge im *Sauromatum*-Kolben. III. *Proceedings K. Nederlandse akademie van weterschappen, 40:* 709–719.

HESS, C. M. & MEEUSE, B. J. D., 1967. The effect of various uncouplers on the respiration of appendix tissue slices of *Sauromatum guttatum* Schott (Araceae) at various stages of anthesis. *Acta botanica neerlandica, 16:* 188–196.

HESS, C. M. & MEEUSE, B. J. D., 1968. Factors contributing to the respiratory flare-up in the appendix of *Sauromatum* (Araceae). I. *Proceedings K. Nederlandse akademie van wettenschappen (Ser. C), 71:* 443–455.

JOHNSON, T. F. & MEEUSE, B. J. D., 1972. The phosphofructokinase of the *Sauromatum*-appendix (Araceae). Purification, and activity-regulation *in vitro*. *Proceedings K. Nederlandse akademie van wetenschappen (Ser. C), 74:* 1–19.

KNOLL, F., 1926. Insekten und Blumen. Experimentelle Arbeiten zur Vertiefung unserer Kenntnisse über die Wechselbeziehungen zwischen Pflanzen und Tieren. IV. Die *Arum*-Blütenstände und ihre Besucher. *Abhandlungen der Zoologisch-botanischen Gesellschaft in Wien, 12:* 379–482.

KNUTSON, R. M., 1972. Temperature measurements of the spadix of *Symplocarpus foetidus* (L.) Nutt. *American Midland Naturalist, 88:* 251–254.

KNUTSON, R. M., 1974. Heat production and temperature regulation in Eastern Skunk Cabbage. *Science, 186:* 746–747.

LAMARCK, J. B. de., 1778. *Flore française, 3:* 1150. Paris.

LANCE, C., 1972. La respiration de l'*Arum maculatum* aucours du développement de l'inflorescence. *Annales des Sciences naturelles, Botanique (12e ser.), 13:* 477–495.

LEICK, E., 1915. Die Erwärmungstypen der Araceen und ihre blütenbiologische Deutung. *Bericht der Deutschen botanischen Gesellschaft, 33:* 518–536.

PRIME, C. T., 1960. *Lords and Ladies*, London: Collins.

RICHARDSON, I, 1966. Studies on the biogenesis of some simple amines and quaternary ammonium compounds in higher plants. Isoamylamine and isobutylamine. *Phytochemistry, 5:* 23–30.

SCHMUCKER, Th., 1925. Beiträge zur Biologie und Physiologie von *Arum maculatum*. *Flora, 118:* 460–475.

SIMON, E. W., 1962. Valine decarboxylation in *Arum* spadix. *Journal of Experimental Botany, 13:* 1–4.

SIMON, E. W. & CHAPMAN, T. A., 1961. The development of mitochondria in *Arum* spadix. *Journal of Experimental Botany*, *12:* 414–420.

SMITH, B. N. & MEEUSE, B. J. D., 1966. Production of volatile amines and skatole at anthesis in some arum lily species. *Plant Physiology*, *41:* 343–347.

SPRENGEL, C. C., 1793. *Das entdeckte Geheimnis der Natur im Bau und in der Befruchtung der Blumen.* Berlin.

VOGEL, S., 1963. Duftdrüsen im Dienste der Bestäubung. *Abhandlungen Mathematisch-naturwissenschaftliche Klasse, Akademie der Wissenschaften und der Literatur, Mainz, 10:* 599–763.

VOGEL, S., 1973. Fungus mimesis of fungusgnat flowers. In N. B. M. Brantjes & H. F. Linstens (Eds), *Pollination and Dispersal:* 13–18. University of Nijmegen: Publ. Dept. Botany.

Insect pollination syndromes in an evolutionary and ecosystemic context

M. C. F. PROCTOR

Department of Biological Sciences, University of Exeter

Evolution of flowers from a (broadly) magnolioid origin can be visualised in terms of three main trends:

(a) Increasingly close adaptation to (and dependence upon) particular specialised flower-visiting insects, exemplified by the specialised zygomorphic flowers. In this group reproductive isolation between related species is often closely involved with the flower-insect relationship, and selection pressures tend to close adaptation and diversity of form and pattern.

(b) Dependence for pollination on repeated visits by a wide range of common and more-or-less promiscuous visitors. In this large and important group, reproductive isolation is commonly maintained physiologically. Such structural specialisation as there is is different in kind, and there are important selection pressures towards uniformity as well as towards diversification.

(c) Return to anemophily, mainly in gregarious plants of open, low-diversity habitats, and cool-temperate forests.

These trends are not sharply demarcated, and can be seen, not as competitive alternatives, but as parts of a pattern of niche differentiation, the balance between them varying from habitat to habitat. The classical specialised insect-pollination syndromes appear as a specialist element within this pattern; some general features of their distribution are briefly examined.

KEY WORDS:—angiosperm evolution—magnolioid—entomophily—parallelism—polyphily—anemophily—ecosystem.

CONTENTS

ORIGINS

In recent years, various lines of evidence and conceptual developments have come together to produce something approaching a consensus view on the main lines of evolution in the angiosperms, and to make possible an integrated evolutionary and ecological view of pollination. The aim of the present paper is to draw together some topics bearing on this field, some familiar, and some less often considered in this context.

Most botanists would probably now agree that the angiosperms have diversified from a (broadly) magnolioid stock, and that the flower-insect relationship is intimately bound up with the origin and primary radiation of the angiosperms

(Baker & Hurd, 1968; Cronquist, 1968). We still cannot pinpoint the origin of the closed carpel in geological time, or say anything with certainty about the immediate precursors of the angiosperms, but these and some other controversial points (Corner, 1949; Cronquist, 1968: 114; Meeuse, 1973) need not greatly affect the central thesis. Palynological evidence (Muller, 1970; Walker & Doyle, 1975; Hughes, 1976) now allows us to say with fair confidence that ancestral angiosperms (or plants close to them) were no more than sparingly represented amongst a predominantly gymnospermous flora, from the late Jurassic through to the earlier part of the Cretaceous period; undoubted angiosperm pollen grains begin to appear in fair numbers and show some incipient diversification in the Albian. Judging from pollen and the first appearance of macrofossils in quantity, the angiosperms expanded rapidly in numbers and diversity from the Cenomanian stage of the Cretaceous onwards, and the major radiation of the group appears to have taken place in the Upper Cretaceous. Those fossil identifications that can be regarded as reliable yield a picture in which, broadly speaking, those families and genera that appear first are among those usually considered the less advanced, while groups that on other grounds are considered highly advanced do not appear until later. The relevant insect groups show a broadly parallel pattern (Harland *et al.*, 1967). The fossil record thus supports the idea of co-evolution of flowers and insects to the extent of providing the right sort of broad correlations, but without providing detailed sequences of instances. This is particularly a limitation of the palynological evidence, which on the other hand is particularly valuable in providing a continuous and semi-quantitative record of the early development and establishment of the angiosperms as the dominant group in the world flora.

From this point, we have to rely upon inference based on present-day morphology and ecology. We can accept that beetles exemplify the unspecialised flower visitors with which the flower-insect relationship originated; cantharophily is common among the Magnoliales at the present day (van der Pijl, 1960; Baker & Hurd, 1968; Thien, 1974). The selective pressure initially responsible for establishing this as a regular association was probably simply for efficiency of cross-pollination. The distribution of pollen by wind is dependent only on the physics of the situation. The concentration of pollen grains falls off rapidly from the source, so for all but gregarious plants wind-pollination is relatively inefficient, and even casual and promiscuous flower visitors may transfer pollen more efficiently among sparsely dispersed flowers. This is well illustrated (though for a specialised insect) by the data of Levin & Berube (1972) on the efficiency of pollen transfer by *Colias*. Further, the pattern of pollen dispersal is at least to some extent responsive to changes in density of the flowers (Levin & Kerster, 1969a, b). Once a flower-insect relation is established, the way is open to selection for increased efficiency and increased specialisation.

A line of development of this sort was visualised by Leppik (1957, 1968, 1971) in his suggested co-evolutionary sequence, from 'haplomorphic' through 'actino-morphic' to 'pleomorphic' flower forms with increasing importance of the recognition of form and pattern by the visiting insects, and then to 'stereomorphic' and 'zygomorphic' flowers associated with the evolution of structural and behavioural characteristics to cope with them in the insect. The earliest part of Leppik's sequence—the evolution of the actinomorphic/pleomorphic flower—is basic to all subsequent angiosperm evolution. It establishes the fundamental insect pollination syndrome: conspicuous perianth members, nectar production, and androecium and gynoecium members in compact spirals or whorls.

SPECIALISED ENTOMOPHILY

The basic selection pressure underlying this development can be seen as increased efficiency of pollination through increased specialisation, the less efficient unspecialised visitors being excluded in the process. In a relationship with bilaterally symmetrical pollinators, the selection pressures towards evolution of zygomorphy are comprehensible enough. Once embarked on an evolutionary course of this sort, an obvious trend is to progressively greater specialisation leading to niche differentiation in relation to particular pollinators (Levin, 1970), resulting in adaptive radiation of a kind common also in other contexts (Janzen, 1973). However, it is clear that efficiency (and certainty) of pollination is far from the only, or even the main, selection pressure operating among specialised stereomorphic and zygomorphic flowers. In saying this I am not thinking primarily of the achievement of outcrossing. The need for *some* outcrossing is axiomatic; how much depends on various ecological considerations, and its regulation can be achieved in various ways, of which flower form is not especially effective (Grant, 1958; Stebbins, 1958). But an important and often dominant factor is that plants with flowers of this kind typically rely at least in part on anthecologcial relationships for reproductive isolation from related species (Grant, 1949). Despite doubts expressed by Levin (1970), this must certainly have been a potent factor in the evolution of floral diversity, as much of the observed 'reproductive character displacement' is between species sharing the same pollinators. Reproductive isolation may be brought about through pollinator relationships in three main ways. Firstly, a group of species may diversify to different pollinators (and this will clearly reinforce the tendency to diversification just indicated). Secondly, pollen transfer between different species may be limited by structural adaptations even though they share the same pollinator. Finally, mechanical isolation of this sort may be supplemented or replaced by features of flower colour, form or scent which lead through the flower-constancy or preferences of the pollinator to ethological isolation (Levin & Schaal, 1970).

An effect of reliance on pollinator relationships for reproductive isolation is that this isolation is rarely perfect. Some gene exchange can therefore take place between plants which may be strikingly different in flower form. This may be not just a way of keeping the evolutionary options open, but an essential characteristic of plants that operate a set of anthecological relationships of this kind. Within any particular evolutionary line we have a complex of selection pressures towards diversity operating in relation to a limited number of groups of specialised pollinators. The consequence is seen in an interplay of adaptive radiation within particular groups of plants and convergent evolution of members of diverse plant families into a limited number of available syndromes of adaptation to particular pollinators (van der Pijl, 1961). The more specialised bee-flower relationships can probably safely be regarded as defining a mainstream in flower evolution, but it is clear from even a cursory examination of zygomorphic bee-pollinated flowers that we are dealing with an evolutionary thicket, not an evolutionary tree, even though the major developments of zygomorphy are concentrated in only three of the ten subclasses of the Takhtajan-Cronquist scheme, the Rosidae, Asteridae and Liliidae. Parallelism is beautifully demonstrated by the clearly independent sternotribic zygomorphy of *Aconitum* (Plate 18), *Corydalis* (both subclass Magnoliidae) and the Leguminosae (Rosidae), and the equally clearly polyphyletic zygomorphy of *Viola* (Dilleniidae), *Impatiens* (Rosidae), Scrophulariaceae and Labiatae (Asteridae), Orchidaceae (Liliidae) and others. Parallelism can be seen again within any of the other major specialised insect pollination syndromes, for instance amongst flowers pollinated by night-flying moths, e.g. *Silene nutans* (Caryophyllidae) (Plate 15A), *Lonicera periclymenum* (Asteridae) (Plate 15B), *Platanthera chlorantha* (Liliidae).

Beautiful examples of adaptive radiation to different pollinators have been described in Polemoniaceae (Grant & Grant, 1965), *Aquilegia* (Grant, 1952, 1976), *Penstemon* (Straw, 1956) and elsewhere. The European orchids provide a striking instance. *Gymnadenia* and *Anacamptis* are pollinated by butterflies and day-flying moths, and *Platanthera* by night-flying moths. *Dactylorhiza* species are probably pollinated mainly by long-tongued Diptera. *Coeloglossum* is apparently adapted mainly to beetles, with *Listera ovata* as a nicely parallel but unrelated case mainly dependent on ichneumons and beetles (Plate 13A, B).

Features leading to mechanical and ethological isolation add a more subtle diversity within the syndromes of adaptation to particular groups of pollinators. Excellent instances of mechanical isolation have been described in *Pedicularis* by Macior (1970, 1973, 1975) and Sprague (1962). We may embrace Stebbins' notion of genetic uniformitarianism (Stebbins, 1972) to conjecture that some of the floral differences between common European genera of Labiatae, which provide good illustrative examples of this kind, may owe their origin to adaptations for mechanical isolation (*Lamium album*, *Ajuga reptans*, *Salvia pratensis*) (Plate 16A, B, C. That the flowers we have been considering typically show striking differences of colour and pattern whenever a potentially flower-constant pollinator is involved needs no emphasis.

Even specialised insect pollination syndromes are not always exclusive or sharply circumscribed, and we should not expect them to be so. Thus *Listera ovata* shows a beautiful precision mechanism built around unspecialised pollinators; the flower is perhaps most characteristically operated by ichneumons, but it is also pollinated effectively and probably commonly by beetles. Many other examples could be adduced, and we can take this as an instance of a general phenomenon. It is easy to be misled by casual and unimportant flower visitors; Grant & Grant (1965) emphasise that pollination may be concentrated into brief periods of the day and of the flowering season. Nevertheless, cases of adaptation to a balance between alternative pollen vectors are certainly common. We should at least be open to the idea that the 'syndromes of adaptation' represent noda in a continuously varying field of adaptation, exemplified with very varying degrees of clarity and exclusiveness by different species.

POLYPHILIC ENTOMOPHILY

The end-products of the lines of development leading to the specialised zygomorphic flowers are so fascinating that it is easy to neglect those flowers, much more numerous in individuals and no less dependent on insects for pollination, whose adaptations are more generalised. For a gregarious plant in a situation where potential pollinators are abundant there is little virtue in pollinator specialisation. Certain, and indeed repeated, visits are assured to a flower which need possess no more than the basic adaptive syndrome for insect visitors. The most striking instances of exploitation of this possibility are the Compositae and Umbelliferae. Some of the essential features of the anthecology of the former were discussed by Burtt (1961), who contrasted the composite capitulum, requiring repeated visits to many small flowers with single ovules, and the large zygomorphic flower in which a single visit may result in the fertilisation of many ovules. One may note in passing that there *are* large zygomorphic flowers with few ovules, notably in the Labiatae. The selective pressures determining ovule number (Johnson & Cook, 1968) do not necessarily run parallel to those determining flower form. Nevertheless, there is no doubt that in general the correlation holds, and that it in part reflects the likely number of other plants available in the neighbourhood as potential pollen sources in the two cases.

Both the Compositae and the Umbelliferae are visited by very large numbers of insects, whether reckoned in individuals or species. Harper & Wood (1957) listed 178 species as visitors to *Senecio jacobaea*; no less than 334 species were recorded visiting carrot flowers at Logan, Utah (Bohart & Nye, 1960). But these are merely extreme instances of a general characteristic of many flowers. Functionally, the composite capitulum is in a sense only a high sophisticated and more finely controlled buttercup. The Ranunculaceae possess the ingredients to make such a flower directly, with achenes individually pollinated through their separate stigmas. A family that starts from a stereomorphic group plan can produce a similar functional unit most readily by reducing flower size and condensing an inflorescence (a well known Englerian trend), a process whose results can be seen in the 'brush-blossom' inflorescences of *Mentha aquatica*, *Armeria maritima*, *Sanguisorba officinalis* and many others. The typical Compositae capitulum is a culmination of this kind of trend, and it may fairly be regarded as highly and specifically adapted to attract and utilise the widest possible range of pollinators— a specialised generalist. The 'zygomorphy' of ligulate florets of Compositae has nothing to do with that of the typical large zygomorphic flowers. Bilateral symmetry has evolved repeatedly; it can evidently arise relatively easily in response to appropriate selection pressures.

Reliance on repeated visits by unspecialised pollinators effectively rules out interspecific isolating mechanisms based on pollinator relationships. In polyphilic flowers, isolation normally depends on physiological cross-incompatibility between related species. Under these conditions there is no special premium on floral diversity. Indeed, it becomes more important that a potential pollinator should recognise a flower as such than that it should distinguish one species from another. Certain flower forms appear in unrelated families in a strikingly repetitive way which can be attributed to Muellerian mimicry; there seems no other good reason for the superficial resemblance between *Ranunculus*, *Potentilla* (Plate 19), *Helianthemum* and small yellow Composite capitula. Other recurring patterns centre around the white and yellow daisy type (*Batrachium*, *Dryas octopetala*, etc.) and the white umbel (*Achillea*, corymbose Cruciferae). That differing reflection of ultra-violet introduces greater diversity to the insect than is apparent to our eyes need not invalidate the general proposition.

ANEMOPHILY

There is of course much pollinator specialisation among plants exemplifying the broad trend we have just considered. In fact the two trends that have been outlined are not sharply demarcated from one another. They are connected by a spectrum of intermediate situations, and each has been at least to some extent recruited from the other. Many cases exist in which the co-adaptative relationship common to these trends has broken down. Nectar-thieving bumblebees biting through the corollas of long-tubed flowers obtain a one-sided advantage (Plate 22B). So do many flowers which depend on sapromyophily or pseudocopulation for pollination. However, in quantitative terms these trends are of minor importance. That of secondary anemophily, is of much greater moment ecologically. It appears, predictably, among those gregarious species (and species of open, exposed habitats) for which wind-pollination is relatively most efficient—the dominant trees of cool-temperate forests, and the dominant species of grasslands and saltmarshes, to name a few instances. Just as no sharp boundary can be drawn between the two major trends in entomophily, so is there no completely sharp line between generalised entomophilous and anemophilous flowers. In particular, brush-blossoms are well preadapted to anemophily and there are a number of instances of species or groups of species which straddle the borderline between insect and wind pollination (e.g. *Thalictrum, Sanguisorba, Castanea, Salix, Plantago*). Which is the predominant mode of pollination in a particular case can only be settled by observation and experiment (Free, 1964; Hatton, 1965; Leereveld, Meeuse & Stelleman, 1976; Stelleman & Meeuse, 1976).

POLLINATION IN THE ECOSYSTEM

The essential point is that these three different patterns of pollination adaptation are not antagonistic alternatives, or stages in an evolutionary sequence one of which is 'better' or 'more advanced' than another. They are three essentially complementary solutions of the problem, each appropriate in a particular range of ecological situations within the ecosystem, and each is a trend within which high specialisation is possible. It is noteworthy that these three lines of adaptation are exemplified by three of the largest and most advanced families of flowering plants, the Orchidaceae, the Compositae and the Gramineae.

Further, these patterns of adaptation must have grown up in a functional relation to one another. They coexist in modern ecosystems, where they are subject to the same sort of competitive and synergistic interactions that underlie the functioning of the ecosystem as a whole. This must always have been so in the past, so we can see the evolutionary process as producing not just diverse lines of floral adaptation in isolation, but interacting and ecosystem-wide patterns of anthecological adaptation. Some features of this sort of interaction are commonplace. There is clearly a selective advantage in a concentrated flowering period for an individual plant species, and for the spacing through the season of the flowering periods of different species (Mosquin, 1971; Heinrich, 1975a, b, 1976b; Reader, 1975; Lack, 1976). Adaptive radiation of flowers to the range of available pollinators (and *vice versa*) is another aspect of ecosystemic interaction. Its spectacular

manifestation is seen in the specialised insect pollination syndromes; it is manifested more subtly in a degree of character displacement between species which
ensures that their effective 'pollen-vector niches' are not identical (Levin, 1970;
Heinrich, 1975a).

Considering the pollination relationship simply in terms of niche differentiation
and the apportionment of pollination resources within the ecosystem, the
specialised entomophilous syndromes represent relatively narrow ecological niches.
For obvious reasons the exploitation of pollination resources roughly parallels
the exploitation of total resources—the biomass (or other measure of 'amount')
of the plant in the community. A very abundant species cannot specialise on a
limited resource for pollination; a rare species may or may not, but is likely to do
so. Moldenke (1975) suggests that in most environments there are selective
pressures towards specialisation, but that in all but the most predictable habitats
pressures favouring behavioural and genetic flexibility militate against this. In
addition there are limitations in the specialisation that the basic architecture of
many flowers can achieve, and there is a density below which a flower will cease to
be energetically rewarding to a flower-constant pollinator (Levin, 1972). The
effects of this may be mitigated by the 'minoring' habit of bumblebees (Free, 1970;
Heinrich, 1975b, c) or by mimicry of other flowers. It is possible that the
orchid pollinium evolved as a device aiding pollen carry-over, thus sidestepping
the need for flower-constancy (Heinrich & Raven, 1972).

Heinrich (1975b) has reviewed energetic aspects of flower-pollinator relationships. From a different (but related) point of view we can look at the distribution
of pollination niches amongst species in terms of apportionment of the available
taxonomic and adaptive range of pollinators in a given habitat. Some relationships
of this kind are readily apparent. Whitehead (1969) detailed reasons for the
prevalence of zoophily in tropical rain forests. Moldenke (1975) has drawn from
both points of view in his valuable study of pollination niche relationships in an
altitudinal transect in California. He found broadly similar frequency distributions
of polyphilic and oligophilic plant species, and of polytropic and oligotropic
flower visitors, in a wide range of communities. In all communities studied, a
large proportion of vectors visited two to five plant species, i.e. were oligotropic
compared with monotropic 'superspecialists' on the one hand or polytropic
'supergeneralists' on the other. Variations in the patterns were more closely related
to community type than to altitude as such, though the rather surprising conclusion emerged that superspecialist and supergeneralist visitors predominated
in the extreme conditions of the high altitude sites. However different constraints
apply to flowers as against visitors, so their niche relationships are not symmetrical. Situations which favour monotropic bees need not favour monophilic
flowers (Cruden, 1972), and as Heinrich (1976a) points out, the specialised social
bees cannot be monotropic because of the long duration of the colony relative
to the flowering seasons of the plants.

PARTITION OF POLLINATION RESOURCES

How do the trends of anthecological adaptation outlined earlier in this paper
relate to the partitioning of pollination resources in natural ecosystems? Experience

in many ecological situations suggests that we should expect a substantial range of niche-widths in all but the most species-poor ecosystems, and that specialist species with narrow ecological niches should be most numerous in high diversity communities. Further, narrow niche specialists should be relatively more abundant in stable (or at least predictable) situations (MacArthur, 1965; Preston, 1962; Southwood, May, Hassell & Conway, 1974). However, a community made up only of narrow specialists with few interactions would be refractory to evolutionary change and vulnerable to perturbations (S. A. Levin, 1970).

Figure 1 shows the distribution of the three broad anthecological trends amongst the plant communities of the Burren district of Co. Clare in western Ireland. The communities are arranged by phytosociological classes in order of 'sociological progression' (Braun-Blanquet, 1964; Lohmeyer *et al.*, 1962). This sets out to place the simplest and least-integrated communities first, leading up to the most complex and highly integrated communities at the end; it can be expected to be free of preconceived anthecological bias. The data were collected by the authors (Ivimey-Cook & Proctor, 1966) within a limited area of country in the space of

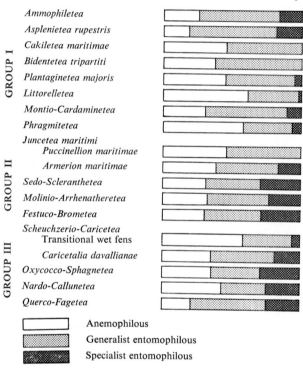

Figure 1. Distribution of major anthecological adaptive types in plant communities of the Burren district, Co. Clare, Ireland, arranged by phytosociological classes in 'sociological progression' (Lohmeyer *et al.*, 1962). Data from Ivimey-Cook & Proctor (1966). The distinction between generalist and specialist species is necessarily arbitrary, but has been kept as consistent as possible. In general, flowers with nectar and pollen concealed within a floral tube, or with other characteristics strongly limiting the range of potential pollinators, have been regarded as 'specialist' and all others as 'generalist'. Compositae–Cynareae have been included with other members of the family in the generalist group. Predominantly autogamous species (a minority in all the communities considered) have been included with their nearest entomophilous relatives. The classes Lemnetea and Zosteretea, very species-poor and with abiotic pollination, are omitted from the diagram.

two summer months in 1959. They too should be free of bias in our context. The increasing proportion of specialised entomophilous species in increasingly species-rich and integrated communities is clearly apparent. If the communities are divided arbitrarily into three successive groups as shown, virtually all the species-rich and stable communities are included in the last group, and some interesting relationships appear. Using the chi-squared test, there is no statistically significant association of the proportions of generalised entomophily and anemophily with sociological progression, and no significant association of the three major anthecological types between the first two phytosociological groups. However, the ratio of entomophily as a whole to anemophily is significantly associated with sociological progression ($P < 0.01$), so this and the strong association of pollination type with community type ($P < 0.001$) is evidently due to the highly significant association ($P < 0.001$) of specialised entomophily with the stable and species-rich communities of Group III.

Some suggestive (but not statistically significant) relationships can be discerned within each of Groups I to III. Group I is obviously heterogeneous, the inland rock-crevice and the mobile dune communities (*Asplenietea* and *Ammophiletea*) having higher proportions of 'specialised' species than the others. Although the difference involves only a few species, it could be argued that these communities have greater permanence and predictability than others of comparable richness, and that the sparse distribution of individuals would militate against anemophily. The similar ratio of anemophilous to entomophilous species in Groups I and II may be partly fortuitous as there are great differences between classes within each group. Those classes typically represented by extensive, dense communities of low diversity have high proportions of anemophilous species. Those composed of communities which are typically fragmentary or linear and which, although they themselves may be species-poor with strong dominance (low alpha diversity in the sense of Whittaker, 1960) normally occur in close proximity to other communities (i.e. in situations of high beta diversity), have high proportions of generalist entomophilous species. It is noteworthy that among both the saltmarshes (*Juncetea*) and the fens (*Scheuchzerio-Caricetea*) the earlier, wetter stages are poorer in specialist species than the latter, drier stages.

There is remarkably little difference in the spectra of the communities in Group III. The grasslands and woods have between 20 and 30% of 'specialist' species; the proportion in the more mature fens is not much lower. The present data are too scanty for firm conclusions about the bogs and heaths, but the figures suggest broadly similar proportions in these too. The species lists from the heaths of the Lizard Peninsula described by Coombe & Frost (1956) give overall proportions of about 28% anemophilous, 42% generalist entomophilous and 30% specialist entomophilous species. The species lists of Pojar (1974) from saltmarsh, bog and alpine meadow communities again suggest similar proportions to ours.

The woods (*Querco-Fagetea*) have the highest proportions of entomophilous species noted here. Are there situations in which higher proportions might be expected in otherwise comparable temperate conditions? An obvious possibility is in the high beta diversity situation of woodland margins—the '*Mantel*' and '*Saum*' communities of the Continental phytosociologists (Tüxen, 1950; Ellenberg, 1963). Görs & Müller (1969) give a synoptic table of a number of south-west

German *Saum*—herbaceous woodland margin—communities. Analysis of those from well-drained sites (*Geo-Allarion*) on the same basis as the Burren communities gives about 20% anemophilous, 58% generalist and 22% specialist entomophilous species, almost the same as the Burren hazel and ash woods. In terms of energetics there may be a much larger difference, as the dominant species of the woodland margin are usually entomophilous. It might be said that what passes for a 'wood' in the Burren is usually little more than a patch of scrub, and as Ellenberg (1963: 692) points out, wood-margin species have become widely distributed through extensive woods as a result of forestry practices. Nevertheless, perusal of other species-lists from forests suggest that the figures we have been considering are not untypical, and that they can be taken as indicating a real and ecologically determined upper limit to the proportion of specialist entomophilous species that a community can support.

This analysis suggests that in terms of total species-composition, there may be less variation from community to community in the proportions of broad anthecological adaptation types than we have tended to think. Comparing temperate and tropical forests, the important difference is in the dominants, between anemophily and entomophily—important in energetic terms, but involving a relatively small proportion of the species. Recent investigations in the Arctic (Hocking, 1968; Kevan, 1972) show that entomophily is very much more important there than some earlier authors have implied. It is only in the most severe, species-poor unstable environments that specialist entomophilous flowers cease to be a significant element in the community.

It would be wrong to imply too regular and tidy a set of relationships amongst patterns of antecological adaptation within ecosystems, and there are plenty of paradoxes and apparent exceptions that spring to mind. But the relationships do seem regular enough to suggest that interesting and potentially illuminating ecological problems may be found in those situations where there are unusual proportions of specialised entomophilous flowers, either in terms of species or of individuals.

REFERENCES

BAKER, H. G. & HURD, P. D., 1968. Intrafloral ecology. *Annual Review of Entomology, 13:* 385–414.
BOHART, G. E. & NYE, W. P., 1960. *Insect Pollinators of Carrots in Utah.* Agric. Exp. Station, Utah State Univ., Logan, Publ. no. 419.
BRAUN-BLANQUET, J., 1964. *Pflanzensoziologie*, 3rd ed. Wien.
BURTT, B. L., 1961. Compositae and the study of functional evolution. *Transactions of the Botanical Society of Edinburgh, 39:* 216–232.
COOMBE, D. E. & FROST, L. C., 1956. The heaths of the Cornish serpentine. *Journal of Ecology, 44:* 226–256.
CORNER, E. J. H., 1949. The durian theory or the origin of the modern tree. *Annals of Botany (NS), 13:* 367–414.
CRONQUIST, A., 1968. *The Evolution and Classification of Fowering Plants.* London: Nelson.
CRUDEN, R. W., 1972. Pollination biology of *Nemophila menziesii* (Hydrophyllaceae) with comments on the evolution of oligolectic bees. *Evolution, 26:* 373–389.
ELLENBERG, H., 1963. *Vegetation Mitteleuropas mit den Alpen.* Stuttgart: Eugen Ulmer.
FREE, J. B., 1964. Comparison of the importance of insect and wind pollination of apple trees. *Nature, 201:* 726–727.
FREE, J. B., 1970. The flower constancy of bumblebees. *Journal of Animal Ecology, 39:* 395–402.
GÖRS, S. & MÜLLER, T., 1969. Beitrag zur Kenntnis der nitrophilen Saumgesellschaften Sudwestdeutschlands. *Mitteilungen der Floristisch-soziologischen Arbeitsgemeinschaft (N.F.), 14:* 153–168.
GRANT, V., 1949. Pollination systems as isolating mechanisms in angiosperms. *Evolution, 3:* 82–97.
GRANT, V., 1952. Isolation and hybridisation between *Aquilegia formosa* and *A. pubescens. Aliso, 2:* 341–360.

GRANT, V., 1958. The regulation of recombination in plants. *Cold Spring Harbor Symposia on Quantitative Biology, 23:* 337–363.

GRANT, V., 1976. Isolation between *Aquilegia formosa* and *A. pubescens*: a reply and reconsideration. *Evolution, 30:* 625–628.

GRANT, V. & GRANT, K. A., 1965. *Flower Pollination in the Phlox Family.* New York: Columbia Univ. Press.

HARLAND, W. B. *et al.,* 1967. *The Fossil Record.* Geological Society of London.

HARPER, J. L. & WOOD, W. A., 1957. Biological Flora of the British Isles. *Senecio jacobaea* L., *Journal of Ecology, 45:* 617–637.

HATTON, R. H. S., 1965. Pollination of mistletoe (*Viscum album* L.). *Proceedings of the Linnean Society of London, 176:* 67–76.

HEINRICH, B., 1975a. Bee flowers: a hypothesis on flower variety and blooming times. *Evolution, 29:* 325–334.

HEINRICH, B., 1975b. Energetics of pollination. *Annual Review of Ecology and Systematics, 6:* 139–170.

HEINRICH, B., 1976a. Resource partitioning among some eusocial insects: bumblebees. *Ecology, 57:* 874–889.

HEINRICH, B., 1976b. Flowering phenologies: bog, woodland, and disturbed habitats. *Ecology, 57:* 890–899.

HEINRICH, B., 1976c. The foraging specializations of individual bumblebees. *Ecological Monographs, 46:* 105–128.

HEINRICH, B. & RAVEN, P. H., 1972. Energetics and pollination ecology. *Science, 176:* 597–602.

HOCKING, B., 1968. Insect-flower associations in the high Arctic with special reference to nectar. *Oikos, 19:* 359–388.

HUGHES, N. F., 1976. *Palaeobiology of Angiosperm Origins.* Cambridge University Press.

IVIMEY-COOK, R. B. & PROCTOR, M. C. F., 1966. The plant communities of the Burren, Co. Clare. *Proceedings of the Royal Irish Academy, 64B:* 211–301.

JANZEN, D. H., 1973. Comments on host specificity of tropical herbivores and its relevance to species richness. In V. H. Heywood (Ed.), *Taxonomy and Ecology:* 201–211. London: Academic Press.

JOHNSON, M. P. & COOK, S. A., 1968. 'Clutch size' in buttercups. *American Naturalist, 102:* 405–411.

KEVAN, P. G., 1972. Insect pollination of high Arctic flowers. *Journal of Ecology, 60:* 831–847.

LACK, A., 1976. Competition for pollinators and evolution in *Centaurea. New Phytologist, 77:* 787–792.

LEEREVELD, H., MEEUSE, A. D. J. & STELLEMAN, P., 1976. Anthecological relations between reputedly anemophilous flowers and syrphid flies. II. *Plantago media* L. *Acta botanica neerlandica, 25:* 205–211.

LEPPIK, E. E., 1957. Evolutionary relationships between entomophilous plants and anthophilous insects. *Evolution, 11:* 466–481.

LEPPIK, E. E., 1968. Directional trend of floral evolution. *Acta Biotheoretica, 18:* 87–102.

LEPPIK, E. E., 1971. Origin and evolution of bilateral symmetry in flowers. *Evolutionary Biology, 5:* 49–85.

LEVIN, D. A., 1970. Reinforcement of reproductive isolation: plant versus animals. *American Naturalist, 104:* 571–581.

LEVIN, D. A., 1972. Low frequency disadvantage in the exploitation of pollinators by corolla variants in *Phlox. American Naturalist, 106:* 453–460.

LEVIN, D. A. & BERUBE, D. E., 1972. *Phlox* and *Colias*: the efficiency of a pollination system. *Evolution, 26:* 242–250.

LEVIN, D. A. & KERSTER, H. W., 1969a. Density-dependent gene dispersal in *Liatris. American Naturalist, 103:* 61–74.

LEVIN, D. A. & KERSTER, H. W., 1969b. The dependence of bee mediated pollen and gene dispersal on plant density. *Evolution, 23:* 560–571.

LEVIN, D. A. & SCHAAL, B. A., 1970. Corolla colour as an inhibitor of interspecific hybridisation in *Phlox. American Naturalist, 104:* 273–283.

LEVIN, S. A., 1970. Community equilibria and stability, and an extension of the competitive exclusion principle. *American Naturalist, 104:* 413–423.

LOHMEYER, W. *et al.,* 1962. Contribution a l'unification du systeme phytosociologique pour l'Europe moyenne et nord-occidentale. *Melhoramento, 15:* 137–151.

MACARTHUR, R. H., 1965. Patterns of species diversity. *Biological Reviews, 40:* 510–533.

MACIOR, L. W., 1970. The pollination ecology of *Pedicularis* in Colorado. *American Journal of Botany, 57:* 716–728 (Lists earlier references.)

MACIOR, L. W., 1973. The pollination ecology of *Pedicularis* on Mount Rainier. *American Journal of Botany, 60:* 863–871.

MACIOR, L. W., 1975. The pollination ecology of *Pedicularis* (Scrophulariaceae) in the Yukon territory. *American Journal of Botany, 62:* 1065–1072.

MEEUSE, A. D. J., 1973. Anthecology, floral morphology and angiosperm evolution. In V. H. Heywood (Ed.), *Taxonomy and Ecology:* 189–211. London: Academic Press.

MOLDENKE, A. R., 1975. Niche specialisation and species diversity along a Californian transect. *Oecologia, 21:* 219–242.

MOSQUIN, T., 1971. Competition for pollinators as a stimulus for the evolution of flowering time. *Oikos, 22:* 398–402.

MULLER, J., 1970. Palynological evidence on early differentiation in angiosperms. *Biological Reviews, 45:* 417–450.

PIJL, L., van der 1960. Ecological aspects of flower evolution. I. Phyletic evolution. *Evolution, 14:* 403–416.

PIJL, L., van der 1961. Ecological aspects of flower evolution. II. Zoophilous flower classes. *Evolution, 15:* 44–59.

POJAR, J., 1974. Reproductive dynamics of four plant communities of southwestern British Colombia. *Canadian Journal of Botany, 52:* 1819–1834.

PRESTON, F. W., 1962. The canonical distribution of commonness and rarity. *Ecology, 43:* 185–215; 410–431.

PROCTOR, M. C. F. & YEO, P. F., 1973. *The Pollination of Flowers.* London: Collins.

READER, R. J., 1975. Competitive relationships of some bog ericads for major insect pollinators. *Canadian Journal of Botany, 53:* 1300–1305.

SOUTHWOOD, T. R. E., MAY, R. M., HASSELL, M. P. & CONWAY, G. R., 1974. Ecological strategies and population parameters. *American Naturalist, 108:* 791–804.

SPRAGUE, E. F., 1962. Pollination and evolution in *Pedicularis* (Scrophulariaceae). *Aliso, 5:* 181–209.

STEBBINS, G. L., 1958. Longevity, habitat and release of variability in higher plants. *Cold Spring Harbor Symposia on Quantitative Biology, 23:* 365–378.

STEBBINS, G. L., 1972. Ecological distribution of centres of major adaptive radiation in angiosperms. In D. H. Valentine (Ed.), *Taxonomy, Phytogeography and Evolution:* 7–34. London: Academic Press.

STELLEMAN, P. & MEEUSE, A. D. J., 1976. Anthecological relations between reputedly anemophilous flowers and syrphid flies. I. The possible role of syrphid flies as pollinators of *Plantago*. *Tijdschrift voor Entomologie, 119:* 15–31.

STRAW, R. M., 1956. Floral isolation in *Penstemon*. *American Naturalist, 90:* 47–53.

THIEN, L. B., 1974. Floral biology of *Magnolia*. *American Journal of Botany, 61:* 1037–1045.

TÜXEN, R., 1950. Grundriss einer Systematik der nitrophilen Unkrautgesellschaften in der Eurosiberischen Region Europas. *Mitteilungen der Floristisch-soziologischen Arbeitsgemeinschaft (N.F.), 2:* 94–175.

WALKER, J. W. & DOYLE, J. A., 1975. The bases of angiosperm phylogeny: palynology. *Annals of Missouri Botanical Garden, 62:* 664–723.

WHITEHEAD, D. R., 1969. Wind pollination in the angiosperms: evolutionary and environmental considerations. *Evolution, 23:* 28–35.

WHITTAKER, R. H., 1960. Vegetation of the Siskiyou Mountains, Oregon and California. *Ecological Monographs, 30:* 279–338.

The pollination of introduced species, with special reference to the British Isles and the genus *Impatiens*

D. H. VALENTINE

Department of Botany, the University, Manchester, U.K.

When an alien species is introduced into a country, and is able to grow and establish itself, its persistence depends on its ability to reproduce. It may do this by vegetative or seed apomixis, or sexually by self- or cross-pollination. Pollination may be by wind or animals, and if the latter, a pollinator has to be found.

Where the invader is not of distant origin, one or more of its natural pollinators may be present, but when it is of distant origin this is less likely. In fact, many introduced species do find a pollinator, doubtless due to the fact that many common pollinators, e.g. species of *Bombus*, are catholic in their preferences, and will seek and use food from flowers wherever it is to be found. Thus, flexibility in the flower-animal relationship, rather than tight adaptation seems often to be mutually advantageous. It is possible that this kind of relaxation of specificity is a feature of disturbed habitats.

From this point of view, species of *Impatiens* in the British Isles are interesting. There is one native and three naturalised species. Of the latter, *I. glandulifera*, of Himalayan origin, is efficiently pollinated by *Bombus* species, as is probably the case in its native area. Another species *I. capensis*, a native of North America, set seeds from cleistogamous flowers, and it is not yet known to what extent its open flowers are visited by insects. In at least part of its native area, it is pollinated by birds.

KEY WORDS:—introduced species—*Impatiens*—*Bombus*—humming birds—*Acaena*—pollination.

CONTENTS

INTRODUCTION

I have for many years been interested in the alien flora of the British Isles. The extent and importance of this flora is probably not generally realised. A recent estimate (which I owe to C. A. Stace) has produced a figure of about 800 naturalised species of seed plants in the British flora. The estimate varies according to the criteria for naturalisation; but the figure is realistic and is on the increase. Questions about the reproductive biology of such species are therefore numerous and of considerable interest; and in many cases, we do not yet know the answers.

When an alien species is introduced into a country, and is able to grow and establish itself, its persistence and spread depends on its ability to reproduce. It

117

may do this by vegetative or seed apomixis, or sexually by self- or cross-pollination. Pollination may be by wind or animals, and if the latter, a pollinator has to be found.

Judging from the statistics produced by Savidge, Heywood & Gordon (1963) for the flora of South Lancashire, most of the species introduced into the British Isles are from Europe, north Africa and west Asia; and when this is so, one or more of the natural pollinators of the plants may be present. But when the introduced species is of more distant origin, this is less likely; and such species have to depend for their pollination on those animals, fortunately not rare, which are catholic in their preferences, and will seek and use food from flowers, wherever it is to be found. This flexibility in the flower-animal relationship is a striking feature of pollination, and as is well known, it is not confined to introduced species.

IMPATIENS

In this contribution, I have chosen to consider the genus *Impatiens*, partly because of personal interest and partly because only one of the species found growing wild in the British Isles is native, the other three being introduced, and having different distributions and modes of pollination. They thus illustrate different aspects of the subject.

The four species, which are all annual herbs, are the native *I. noli-tangere* L. and the introduced *I. glandulifera* Royle, *I. capensis* Meerburgh and *I. parviflora* DC. I shall begin with a brief account of the native species.

I. noli-tangere is widespread in Europe and Asia, where it occurs in moist, shady places; and it extends to western North America. It has typical zygomorphic *Impatiens* flowers which are yellow in colour, and about 3 cm long. They have a large posterior spurred sepal which produces nectar; and they are protandrous, the male stage lasting for about two days. When the androecium is shed, the flower enters the female stage for a few hours, after which the perianth falls. Little is known about the pollinators of this species in the British Isles, but in Czechoslovakia, where it has recently been studied by Daumann (1967), its pollinators are mainly *Bombus* species (Table 1) and this agrees with the observations of Knuth (1908). According to Daumann, insect visitors are scarce, and some seed at least is set in small cleistogamous flowers; but the relative amounts of seed produced by the two types of flowers are not known.

I. glandulifera, the Indian Balsam, the first and the most abundant of the introduced species, is a native of the Himalayas, introduced into England as a garden plant in 1839 (Coombe, 1956b). It was first recorded in the wild in 1848 (Coombe, 1956b), since when it has spread to most parts of the lowlands of the British Isles. In more recent years it has become widespread on the continent of Europe. It is usually found along the banks of rivers and streams, where it may form extensive linear communities. The plant is tall (up to 2 m) and bears inflorescences of large, wine-red zygomorphic flowers with a distinct odour and 3–4 cm long.

The pollination mechanism is similar to that in *I. noli-tangere*, the flower being protandrous. The insect lands at the front of the flower on two large 'petals' (each representing a pair of fused petals), and passes through the corolla to the spurred sepal behind from which it obtains nectar and in which it almost disappears. In so doing, its back is brought into contact with the androecium; in this, the five stamens are joined together laterally into a staminal cone (formed from the filaments) and this releases the sticky pollen in a downwards direction. The insect comes into contact a second time as it withdraws. At this, the male stage, the staminal cone completely encloses the developing ovary.

As the male stage comes to an end, the staminal cone shrivels and falls off, being pushed to some extent by the growth of the ovary, and the ovary with its stigma is exposed. The arms of the stigma then open slightly and it becomes receptive, so that when the next insect visits the flower, which is still producing nectar, its back comes into contact with the stigma, and cross-pollination may occur. Finally the perianth falls off and insects visits cease.

The male stage of the flower is considerably longer than the female stage. Daumann estimates that the life of the flower is two to three days, and that the duration of the female stage is only three to five hours. Our own estimate, made on garden populations, is that the duration of the female stage is about one-fifth that of the male, which is rather longer than Daumann's estimate. In spite of the brevity of the female stage, a high proportion of the flowers usually sets seed. This is probably due to the fact that, in many of its localities, *I. glandulifera* produces a great deal of nectar and is constantly visited by insects. Daumann showed experimentally that insects were attracted visually to the flowers, rather than by odour.

Table 1. Some pollinators of two species of *Impatiens*

| | *I. glandulifera* | |
I. noli-tangere	Britain	Czechoslovakia
—	*Vespa vulgaris*	—
Apis mellifera	*Apis mellifera*	*Apis mellifera*
Bombus terrestris	—	*Bombus terrestris*
B. agrorum	*B. agrorum*	*B. agrorum*
—	*B. lucorum*	*B. lucorum*
—	*B. hortorum*	—
—	—	*B. lapidarius*
—	—	*B. silvorum*

Lists of pollinators are given in Table 1. The list for Britain is a short one, and refers only to some investigations in the area of Manchester (Valentine, 1971). More study would certainly extend the list. In the two semi-natural habitats investigated (in August) wasps were frequent visitors, with bumblebees the next most frequent (Plate 23). In Czechoslovakia, Daumann found that honey-bees were the most frequent, and the plant is apparently cultivated in a few places as a plentiful source of nectar. It is known to be visited in one of its native habitats in the Himalayas by bumblebees, but it has not yet been possible to discover which species.

I. glandulifera is self-compatible, and pollen is often transferred geitonogamously as there may be many open flowers on a single plant, but cross-pollination is obviously very frequent. It may be added that occasional visits to flowers by unidentified, small Diptera were observed in the field; and when flies were admitted by mistake into an insect-proof greenhouse in which plants were growing, some uncontrolled seed-set occurred. But extensive observation in the field indicates that Diptera play only a very minor part. Pollination can also occur by the activity of thrips, or, in the greenhouse, red spider. As regards night visitors, bees and wasps continue their activities far into the dusk; and T. C. Dunn has recently informed me that in Durham, the flowers are often visited at night by noctuid moths. More information about this is desirable.

It is interesting to note the presence of stalked glands at the nodes of *I. glandulifera*. These have been interpreted as extra-floral nectaries. In Britain they rarely, if ever, secrete nectar and do not attract insects, and this agrees with the experience of Daumann in Czechoslovakia.

In migrating (with the help of man) across the Northern Hemisphere from montane meadows in the Himalayas to the European lowlands, *I. glandulifera* has made a move, which in many of its features is paralleled by other species. It is however exceptional in two ways. First, relatively few of the introduced species of western Europe come from India; and secondly, not many are insect-pollinated annuals. Generally speaking the migrants, if annuals, are self- or wind-pollinated; and if insect pollinated, they are perennials.

We turn now to our second introduced species, *I. capensis*. This is a North American species, found mainly in the central and eastern parts of the U.S.A. and Canada; it is similar to *I. noli-tangere* but slightly larger and with flowers which are light orange in colour and 2–3 cm long. It came into Europe as a garden escape and is naturalised on river banks in France and in Southern England, where it was first noted in 1822 (Williams, 1912). The only account of it which we know is a paper by Bennett (1873) who studied a population along the banks of the River Wey. He found that the cleistogamous flowers outnumbered the open flowers very considerably, and the two types of flower were sometimes on separate plants. Seed was set early in the season from cleistogamous flowers: but Bennett never saw insect visitors on the open flowers, which appeared later, though Darwin, in a letter quoted by Bennett, "was almost certain that they are frequently visited by bumble-bees".

This is a remarkable account, and it is much in need of confirmation. The only other British observations known to me at present are first, that made by Coombe (1956b), who records that a geometrid moth (*Xanthorhoe biriviata*) visited *I. capensis* in the south of England; and secondly a note by Lousley (1976) in his *Flora of Surrey*, that the flowers on the plants of this species on Wimbledon Common are always cleistogamous. In north America, there is some positive information that in at least part of its area, *I. capensis* (known as Jewel-weed) is visited by birds and is presumably bird-pollinated. An account by Pickens (1944) makes it quite clear that in a certain area in Kentucky, in the summer and autumn, colonies of *I. capensis* along a small stream are constantly visited by the ruby-throated humming-bird *Archilochus colubris*, the plant colonies being the basis of the territories occupied by individual birds. *I. capensis* has also been observed recently in Virginia by St. Vogel, and he informs me that the flowers there are commonly visited by humming-birds. However, the species extends northwards to Ontario, and here, humming-birds are comparatively rare; and I am informed by J. K. Morton that visits by birds to *I. capensis* have not been observed, though whether insects take their place is not clear. The species occurs also in Newfoundland, where humming-birds are absent; here it must be pollinated by insects, or form its seed by self-pollination, or from cleistogamous flowers.

It is of interest to mention here the case of another introduced species which has some analogies with *I. capensis*. *Fuchsia magellanica* Lam. a native of south America, is widely naturalised in Ireland and Western Britain. In south America, *F. magellanica* occurs from Cape Horn in the south to the Atacama desert in the north. According to D. M. Moore, this range is completely covered by that of the Green-backed firecrown humming-bird (*Sephanoides sephanoides*); and this bird visits the *Fuchsia* frequently in some areas, and must be presumed to be an important if not the main pollinator. In Ireland, according to D. A. Webb, the plant is much visited by hive-bees and is regarded as a good honey-plant; it is doubtless visited by other bees as well. Much of the *Fuchsia* population in Ireland is found in hedges, where it is planted, and the commonest form, which may be triploid, rarely sets seed; but another variety, which is less common, sets seed freely.

It would thus appear that in south America, the *Fuchsia* is largely bird-pollinated, and in the British Isles, insect-pollinated; a similar difference in mode of pollination is likely in *Impatiens capensis*, though here the nature of one of the alternative pollinators is not yet clear. It is at first sight surprising to a European botanist that such a great variation in the nature of the pollen-vector should be tolerated; but this kind of variation is well known, e.g. in California. Thus, as Chase & Raven (1975) have shown, the flowers of two distinct species of *Aquilegia*, formerly thought to be isolated by the preferences of different pollinators, are both visited by humming birds, bumblebees, hawkmoths and flies; and their isolation is conditioned, not only by the pollinators, but by habitat factors as well.

Our third naturalised species is *I. parviflora*. This is a native of central Asia which is now widely naturalised in eastern, central and northern Europe. It was first noticed in Britain in 1851 (Williams, 1912). It occurs on a wide range of well-aerated moist soils and does not often come into contact with the other three species. It is a relatively small plant, and it has pale yellow, small flowers with a straight spur about 1 cm long. It resembles the other species in the general features of its floral mechanism. It is protandrous; the flowers open and the anthers dehisce on one day, the androecium is shed on the following day and the perianth on the next. It is self-compatible, and does not produce cleistogamous flowers. It is pollinated, both in Britain and Czechoslovakia, by small to medium-sized hover-flies, which both suck the nectar and devour the pollen. If the lists of insect visitors given by Coombe (1956a) and Daumann (1967) are compared, it is found that one species, *Syrphus vitripennis*, is recorded from both Britain and Czechosolvakia, and that the lists have three genera in common. Unfortunately, nothing is known of the floral biology of the species in the areas of central Asia in which it is native, not even whether it is pollinated mainly by Diptera. But of its biological and ecological success in Europe there can be no doubt; and this must be due in part to the fact that numerous European species have been able to act as pollinators. In both western and central Europe, it sets plentiful seed.

I think that this very limited survey demonstrates an important evolutionary point; and this is, that both the plants and the animals are very adaptable. Thus the plants, arriving in a completely new environment, have been able to attract the services of insects of unfamiliar species and even genera, and, in the case of *I. capensis* (probably), insects instead of birds (this is definitely the case in *Fuchsia*).

Similarly the insects readily visit flowers which are not native. It is well known that many pollinating insects are polytropic and are not limited to a single species or even genus; but in the case of aliens, completely new flowers are presented, and apparently accepted without difficulty. Evidence that this is also true for introduced species in north America has been given by Mulligan (1972). I might add here some interesting new evidence on this point, kindly made available to me by S. R. J. Woodell (1977), who has recently studied the pollination of introduced species of flowering plants on the oceanic island of Aldabra. He was able to show that many of the introduced species were pollinated by a native sun-bird and a native beetle. The latter was seen to visit more than half of all those species in flower in early 1974; and it was apparently indifferent to the colour or morphology of the flower, or as to whether the species was native or introduced.

The three introduced *Impatiens* species form a group of self-compatible, animal-pollinated annuals, two of which at least are regularly cross-pollinated. Introduced species of this kind are not very common; and this suggests that it would be rewarding to analyse all the introduced species, say, of the British flora from the point of view of their habit, mode of pollination and breeding-system.

It is however of interest to state that so far as is known, and apart from a few dioecious species, there are few records of an introduced species failing to set seed because of the lack of a suitable pollen vector. The only well-authenticated cases are in the Leguminosae. Bumblebees had to be introduced into New Zealand as pollinators for *Trifolium pratense* L., as otherwise the clover could not set seed; and *Melilotus* spp., introduced into north America sometimes fail to set seed for lack of a pollinator (Faegri & van der Pijl, 1966).

CONCLUSION

I hope this paper will have shown that the study of the pollination of introduced species is of interest. It is certainly of importance, in that it gives us vital information about the breeding system and hence about variation in the species concerned. Yet there is little recent work on these lines. To illustrate the gaps in our knowledge, the case may be cited of the southern hemisphere species *Acaena novae-zelandiae* Link. which is naturalised on the Northumberland coast. Current floras provide no information on pollination; but D. M. Moore, who is familiar with the genus in South America, was able to tell me that in cultivation this and other species of the genus are autogamous, and that there is evidence that in the field some species are wind-pollinated. Observations on the species in Britain have yet to be made.

Such examples are not hard to find; and it is clear that there is an opportunity here for naturalists, with moderate expertise in plants and insects, and time to spend on field observation, to make important contributions to knowledge.

ACKNOWLEDGEMENTS

I am grateful to D. M. Moore, J. K. Morton, D. A. Webb, S. Vogel, C.A. Stace and T. C. Dunn for information; also to Mrs I. Dingwall for many valuable observations.

REFERENCES

BENNETT, A. W., 1873. On the floral structure of *Impatiens fulva* Nuttall with special reference to the imperfect self-fertilised flowers. *Journal of the Linnean Society of London (Botany), 13:* 147–153.

CHASE, V. C. & RAVEN, P. H., 1975. Evolutionary and ecological relationships between *Aquilegia formosa* and *A. pubescens* (Ranunculaceae), two perennial plants. *Evolution, 29:* 474–486.

COOMBE, D. E., 1956a. Biological Flora of the British Isles. *Impatiens parviflora* DC. *Journal of Ecology, 44:* 701–713.

COOMBE, D. E., 1956b. Notes on some British plants seen in Austria. *Veröffentlichungen des Geobotanischen Instituts, Eidgenössiche Technische Hochschule Rübel in Zürich, 35:* 128–137.

DAUMANN, E., 1967. Zur Bestaübungs- und Verbreitungsökologie dreier *Impatiens*-Arten. *Preslia, 39:* 43–58.

FAEGRI, K. & van der PIJL, L., 1966. *The Principles of Pollination Ecology.* Oxford: Pergamon.

KNUTH, P., 1906–9. *Handbook of Flower Pollination.* (Transl. by J. R. Ainsworth Davis). 3 vols. Oxford.

LOUSLEY, J. E., 1976. *Flora of Surrey.* Newton Abbot: David and Charles.

MULLIGAN, G. A., 1972. Autogamy, allogamy and pollination in some Canadian weeds. *Canadian Journal of Botany, 50:* 1767–1771.

PICKENS, A. L., 1944. Seasonal territory studies of Ruby-throats. *Auk, 61:* 88–92.

SAVIDGE, J. P., HEYWOOD, V. H. & GORDON, V., 1963. *Travis's Flora of South Lancashire.* Liverpool Botanical Society.

VALENTINE, D. H., 1971. Flower-colour polymorphism in *Impatiens glandulifera* Royle. *Boissiera, 19:* 339–343.

WILLIAMS, F. N., 1912. *Prodromus Florae Britannicae*, part 9: 503.

WOODELL, S. R. J., 1977. The role of unspecialised pollinators in the reproductive success of Aldabran plants. *Philosophical Transactions of the Royal Society (Ser. B).*

Insect visiting of two subspecies of *Nigella arvensis* under adverse seaside conditions

D. EISIKOWITCH

Department of Botany, University of Tel-Aviv, Israel

Effective pollination by insects of some of the coastal plants of Israel that are exposed to windy conditions is attained by shifting the blooming time in the summer toward midnight and the early morning hours.

Such plants as *Nigella arvensis* however, counteract the effect of strong winds by dwarfism and flower stalk rigidity and this phenomenon has been found to be crucial for the safe and stable landing of the pollinating bees on the flower.

KEY WORDS—Israel—*Nigella*—coastal—wind—dwarf—pollination

CONTENTS

INTRODUCTION

From the standpoint of plant pollination, the sea shore habitats of Israel are problematic, especially because of the inhibiting effect of winds on the winged insect pollinators, and serious damage to the flowers that may be incurred by sea spray (Eisikowitch, 1973).

Some coastal plants (*Pancratium maritimum* and *Oenothera drummondii*) counteract the effect of wind by flowering during mid-summer nights, when the wind is usually low, and are pollinated by hawkmoths.

Others (*Ipomoea stolonifera*) open only during the early morning hours of the summer. Behaviour of these, and of apomictic plants such as *Limonium oleifolium* (Dvoskin, 1969), is aimed at mitigating unfavourable environmental conditions. We deemed it of interest, therefore, to investigate the pollination system of coastal plants which do not "behave" according to the usual "rules", and flower at midday, remain open for a long period, and are pollinated by diurnal insects. For this purpose, two related subspecies of *Nigella arvensis* were chosen.

According to Strid (1969, 1970), *Nigella arvensis* is a widespread species, occurring throughout south and central Europe, south-west Asia and north Africa. In *Nigella*

125

arvensis, as in other species of that genus, development of the flower is characterised by successive spreading of the stamens. The anthers are extrorse and shed their pollen when the stamens form a 45° angle to the floral axis. The styles are erect when young but become excurved as floral development proceeds. As the pollinating insect alights on the petaloid nectaries, it either touches the open stamen or deposits pollen grains on the stigma.

Zohary (1966) investigated this species in Israel and described three subspecies and seven varieties. The two subspecies selected for the present study were: 1, *Nigella arvensis* ssp. *tuberculata* var. *submutica* Brnm.; 2, *Nigella arvensis* ssp. *divaricata* var. *palaestina* Zoh. These two annual subspecies occur within the light soil belt of the coastal plain of Israel. Ssp. *divaricata* populates the high coastal hills which are exposed to wind and spray from the sea, while ssp. *tuberculata* grows further inland, on the low-lying sandy loam plains that are well protected from wind and direct sea spray (Waisel, 1959).

The two subspecies bloom simultaneously from May to mid-July, the flowers opening for about ten days. Inasmuch as no spontaneous self pollination occurs, insects are indispensable for seed formation. Comparison of the flowers of the two subspecies revealed differences in size and shape of the petals and different shapes and patterning of the nectaries (Plate 6A). Photographs made through UV filters failed to show any differences between the flowers of each with respect to UV reflection (Plate 6B).

Plants of ssp. *divaricata* are prostrate, the stems and flower stalks being thick and stout; specimens of ssp. *tuberculata*, in comparison, are thinner, softer and taller (Fig. 1). Ssp *divaricata* has a significantly greater number of open flowers per plant

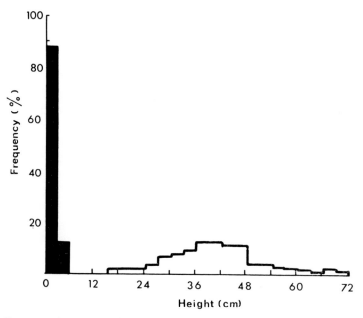

Figure 1. Frequency of occurrence of different heights of *Nigella arvensis* ssp. *divaricata* (left) and ssp. *tuberculata* (right) in the wild populations studied.

Table 1

	ssp. *tuberculata*	ssp. *divaricata*
Mean height of open flowers above ground (cm)	35.64 ± 7.65	1.69 ± 0.54
Mean number of open flowers per plant	1.67 ± 0.85	5.03 ± 1.379
Mean distance between flowers within a plant (cm)	15.90 ± 7.24	4.20 ± 1.48

and the distances between the open flowers are considerably greater in ssp. *tuberculata* than in ssp. *divaricata* (Table 1).

Observations *in situ* on ssp. *divaricata* were made at Tel-Baruch (near Tel Aviv), on a high coastal hill. The ssp. *tuberculata* was similarly studied *in situ* at Hod-Hasharon, in a scattered Eucalyptus wood. In addition, seeds from the two populations were collected for experimental purposes and brought to the Botanical Garden of the Tel Aviv University.

EXPERIMENTS AND RESULTS

Seeds of both subspecies from these localities were sown in pots in the Botanical Garden of the Tel Aviv University. The plants at the natural habitats were left undisturbed until flowering time, whereas those grown in pots were thinned so that each pot contained only one healthy plant.

Some of the pots at the Botanical Garden were left exposed, while others were shaded by various heights of plastic nets.

All plants grown in shaded pots in the Botanical Gardens developed properly, but judging by flower height, the ssp. *tuberculata* responded more strongly to shade from a net than did ssp. *divaricata* (Fig. 2).

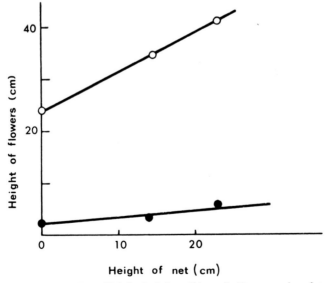

Figure 2. Growth response under artificially shaded conditions. O, *N. a.* ssp. *tuberculata;* ●, *N. a.* ssp. *divaricata.*

Pollinator behaviour

Honey bees comprised the majority of pollinators on ssp. *tuberculata* at Hod Hasharon, with flower visits occurring from 06.00 to 18.00 hrs. Here the wind speed did not exceed 0.5 m/sec and the bees were seen to browse systematically and fly from flower to flower. Specimens of ssp. *divaricata* at Tel-Baruch were also visited by honey bees, but in this case the bees flew very close to the ground and when landing to gather food, they commenced to walk or fly from flower to flower. When the wind speed increased, these bees occasionally encountered difficulty in flying and were forced to walk on the ground in order to reach the flowers. Under extreme wind conditions, the bees could not fly at all and they left the area. In a random mixed plot of the two subspecies at the Botanical Garden, honey bees were again the predominant pollinator, with a few small bees and some wasps also participating. It was found that none of the pollinators could distinguish between the two different subspecies but rather, they landed and gathered food indiscriminately from any flower encountered (Fig. 3).

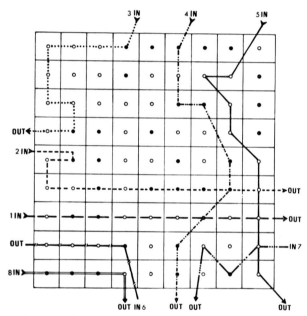

Figure 3. Sequence of nectar-gathering by pollinating insects in a randomised plot at the Botanic Garden. O, *Nigella arvensis* ssp. *tuberculata;* ●, ssp. *divaricata.* Pollinators: 1–5, honey bees; 6, small bees; 7, small wasps; 8, small bees.

Bee flight and wind velocity

Intensive observations carried out during the flowering period of the exposed plants at Tel-Baruch showed that excessive winds force the bees away. In the area investigated, numerous specimens of *Limonium oleifolium* are in full flower during the summer and are then extremely attractive to bees. We found it possible to get very close to the bees during their nectar-collecting and, concurrently, also measure the wind velocity with the aid of an anemometer probe. Correlation of the wind velocity with bee activity gives a fair estimate of the ability of bees to forage under

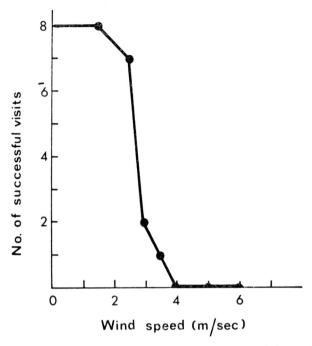

Figure 4. Flower visits by honey bees under windy conditions at Tel-Baruch (on ssp. *divaricata*).

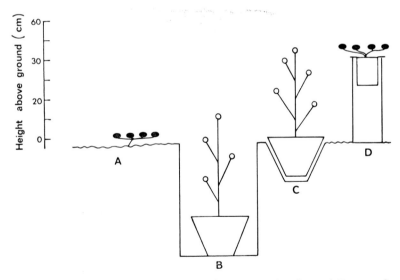

Figure 5. Scheme of the arrangement of transplanted *N. a.* ssp. *tuberculata* and *N. a.* ssp. *divaricata* plants within the exposed area at Tel-Baruch. A. Indigenous plants (ssp. *divaricata*). B. Buried pots of ssp. *tuberculata* with their flowers protruding up to the height of those of ssp. *divaricata*. C. Pots of ssp. *tuberculata* plants at the natural height above ground. D. Pots of ssp. *divaricata* plants artificially raised (on plastic stands) to the level of ssp. *tuberculata* plants. An anemometer probe was placed near each flower.

windy conditions. Thus we found that the bees can reach the flowers at wind veloci-
ties up to 3.0 m/sec, but at 4.0 m/sec wind velocity they are blown away (Fig. 4).

Transplant experiments

Twenty pots of ssp. *tuberculata* plants and ten pots of ssp. *divaricata* plants were
transferred to the exposed area at Tel-Baruch and arranged there according to the
scheme shown in Fig. 5.

As expected from the correlation of wind velocity with plant height above ground,
the bees foraged only on the prostrate plants (Fig. 5, example A), their foraging
ability being limited by wind to heights of 0–5 cm.

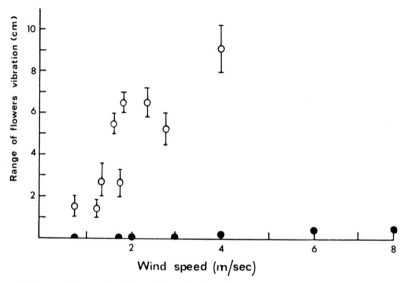

Figure 6. Range of flower vibration under a windy regime. ○, *N. a.* ssp. *tuberculata*; ●, *N. a.* ssp.
divaricata.

Vibration of the plants, a phenomenon not prominent in the inland site of Hod
Hasharon, figured importantly near the sea. Thus plants of the ssp. *tuberculata*
vibrated considerably in almost any kind of wind, and bees invariably foraged on
prostrate plants which were exposed to the least wind and hardly vibrated at all
(Fig. 6). Thus, sunken plants of ssp. *tuberculata* (example B) were not visited,
apparently due to this vibration.

By the end of three days of observations, plants of ssp. *tuberculata* were badly
damaged by salt spray, but the artificially elevated plants of ssp. *divaricata* (Fig. 5,
example D) showed no more damage than the native plants.

DISCUSSION

Distinctions between the two subspecies in flowers, leaves and height, are not
sufficiently great to be detected by bees and other insect pollinators, which visit
both subspecies indiscriminately at the Botanical Garden. Under natural conditions,
bees remain constant to their usual type, as confirmed by Levin & Kerster (1973),

who concluded that "bees tend to remain at a given height on inter-plant flight".

Conditions at the investigated inland area (Hod Hasharon) allow a greater density and cover of vegetation. Here wind velocity does not act as a limiting factor in insect nectar-gathering, but rather the height of the flower above ground is the determining factor. Inasmuch as plants grown in the shade attain greater height (Fig. 2), the advantage of such tall plants becomes obvious. However, in order to reach elevated flowers, the insect must rely strictly on its flying ability, as walking from flower to flower becomes unfeasible.

On the other hand, at the area exposed to the sea (Tel-Baruch), the pollinating insect must overcome two main difficulties. Firstly, it must tackle the problem of the wind, which it solves partly by flying low, and secondly, it must be able to land and fasten on the flower and in this it is partly aided by the stoutness and rigidity of the local subspecies; the stouter and more rigid the flowers, the less they vibrate and the easier the landing, and a relatively great number of open flowers means less search time and energy expenditure per unit food gathered in difficult conditions (Heinrich & Raven, 1972). The nature of the landing site, governed by several characters, can therefore be a target for selection under certain conditions. A similar case has been described by Eisikowitch & Woodell (1975), who showed that bumblebee behaviour appears to be affected by the bending of inflorescence stalks in salt-marsh *Armeria maritima* in Britain.

In conclusion, effective pollination under extreme environmental conditions can be attained by several means; by a shift of the flowering time (Eisikowitch, 1973), by selfing (Baker, 1966; Strid, 1969; Uphof, 1938), or by replacement of the sensitive pollinator by a less sensitive one (Cruden, 1972). However, for an entomophilous out-breeder growing under windy conditions, dwarfism can be a fair solution for pollination under extremely windy conditions, even at diurnal times.

ACKNOWLEDGEMENT

Thanks are due to Mr Y. Natani for his technical assistance.

REFERENCES

BAKER, H. G., 1966. The evolution, functioning and breakdown of heteromorphic incompatibility systems. I, The Plumbaginaceae. *Evolution, 20:* 349–368.

CRUDEN, R. W., 1972. Pollinators in high-elevation ecosystems: relative effectiveness of birds and bees. *Science, 176:* 1439–1440.

DVOSKIN, S., 1969. *Reproductive systems in some Plumbaginaceae in Israel.* M.Sc. thesis, Tel Aviv University.

EISIKOWITCH, D., 1973. Mode of pollination as a consequence of ecological factors. In V. H. Heywood (Ed.), *Taxonomy and Ecology.* London: Academic Press.

EISIKOWITCH, D. & WOODELL, S. R. J., 1975. Some aspects of pollination ecology of *Armeria maritima* (Mill.). Willd. in Britain. *New Phytologist, 74:* 307–322.

HEINRICH, B. & RAVEN, P. H., 1972. Energetics and pollination ecology. *Science, 176:* 597–602.

LEVIN, D. A. & ANDERSON, W. W., 1970. Competition for pollinators between simultaneously flowering species. *American Naturalist, 104:* 455–467.

LEVIN, D. A. & KERSTER, H. W., 1973. Assortative pollination for stature in *Lythrum salicaria. Evolution, 27:* 144–152.

STRID, A., 1969. Evolutionary trends in breeding system of *Nigella* (Ranunculaceae). *Botaniska notiser, 122:* 380–397.

STRID, A., 1970. Studies in the Aegean flora. XVI. Biosystematics of the *Nigella arvensis* complex. *Opera botanica, 28:* 1–169.

UPHOF, J. C. Th., 1938. Cleistogamic flowers. *Botanical Review, 4:* 21–50.

WAISEL, Y., 1959. Ecotypic variation in *Nigella arvensis* L., *Evolution, 13:* 469–475.

ZOHARY, M., 1966. *Flora Palaestina, I.* Israel Academy of Science and Humanities.

EXPLANATION OF PLATE

PLATE 6

A. Sepals and nectaries of *Nigella arvensis* ssp. *tuberculata* (left) and ssp. *divaricata* (right).
B. Same as A, but photographed through ultraviolet filter.

PLATE 6

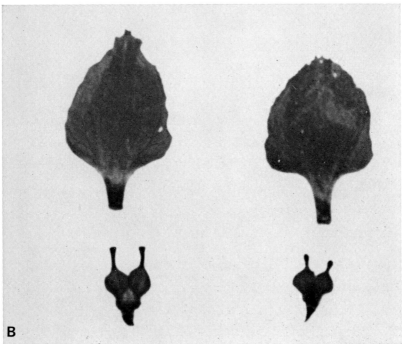

D. EISIKOWITCH

(*Facing page 132*)

Pollinator behaviour and the breeding structure of plant populations

D. A. LEVIN

Department of Botany, The University of Texas, Austin, Texas, U.S.A.

The foraging behaviour of pollinators is a function of the resource quality which is expressed in terms of plant density and quality of reward per plant. The observed foraging behaviour of pollinators is compared with that predicted for optimal foraging strategy, and the implications of this behaviour on the breeding structure of populations is considered. Pollinator behaviour also is considered as it relates to the incidence of interspecific hybridisation in patchy environments, transient environments, and on islands.

KEY WORDS:—foraging — behaviour — pollinators — patch — neighbourhood — hybridisation—specificity—constancy.

CONTENTS

Editor's note: Professor Levin was, through no fault of his own, sadly unable to attend the conference. However, he was most keen that this paper should be presented to the conference, and published. Thus, it was read to the participants, and appears here.

INTRODUCTION

The prime foci in pollination biology have been the description of pollen vectors and their foraging behaviour within and among plants, and of the myriad of morphological, chemical and developmental adaptations and permutations by which plants attract and manipulate pollinators so that they carry pollen from a plant to another of the same species. These topics have been reviewed within historical and conceptual frameworks by Baker & Hurd (1968), Free (1970). Faegri & van der Pijl (1971), Proctor & Yeo (1973) and Heinrich (1975). In spite of considerable interest in the reciprocity between plants and pollinators, little attention has been given to the importance of pollinators to plants except as pollen vectors. The consequences of pollinator foraging behaviour on the breeding structure of populations has been largely ignored by plant evolutionists, although plant breeders long have been aware of these consequences (Levin & Kerster, 1974). The implications of pollinator behaviour in terms of the ethological isolation of plant species have also been appreciated (Grant, 1949).

Progress in pollination biology was paralleled by increased knowledge of breeding systems, gametophytic selection, and levels of self-fertilisation in natural populations. Until recently most estimates of selfing in nature were upon wind-pollinated plants, and failed to take into account the role of mating between relatives in contributing to overall inbreeding (Allard, Jain & Workman, 1968; Jain, 1976). Studies involving animal-pollinated species tacitly assumed that outcrossing was at random.

During the past few years, the "barrier" between research on pollination biology and breeding systems has been "breached". It is now apparent that pollinator foraging behaviour may have a profound effect on the breeding structure of populations because neighbouring plants are likely to be close relatives. More-over, pollinators are much more sensitive to spatio-temporal fluctuations in plant sociology and site conditions than are the breeding systems or the more inclusive genetic system. They may alter the level of inbreeding in a population by an order of magnitude.

My purpose is to explore the genetic implications of pollinator foraging behaviour which may be best understood from the vantage point of the theory of optimal foraging. Foraging theory will be discussed in light of empirical data on pollinator behaviour, and then will be used to generate expectations about pollen flow within and between populations of single species, and between different species.

THE FORAGING STRATEGIES OF POLLINATORS

Pollinators encounter a vast array of potential food plants which vary manifestly in their density, nutritional value, and ease of handling. Faced with this diversity, a pollinator must decide (1) where to search, (2) which species to feed from, and (3) which plants to feed from and in what sequence. The first two decisions are simple. Foraging should be concentrated in habitats where the expectation of yield is the greatest, and upon those plants whose pollen and/or nectar are most efficiently harvested and provide necessary nutrients as well as calories. A plant species should be exploited only if the amount of time spent in travel and extraction is more productive than the same time devoted to another species in the same habitat. The decision to visit a particular plant is more difficult than the previous two, since the pollinator may have no way of knowing the quantity of nutritional reward which awaits it (MacArthur, 1972).

In more specific terms, natural selection favours individuals that are genetically predisposed to forage in a manner that maximises efficiency in feeding and correlatively the number of its offspring. Selection for efficiency in feeding favours increased ability to choose the best plants (with respect to their effects on fitness), and to locate the plant and manipulate the flowers (Emlen, 1973). Optimal foraging strategies in pollinators are expected to parallel those of predators, because the ecological economics of foraging in the two groups are similar, and pollination fits within the broad conceptual framework of predation (Covich, 1974). Optimal foraging strategies for predators have been discussed from different vantage points by several authors (Emlen, 1968; Royama, 1970; Schoener, 1969, 1971, 1974a, b; Rapport, 1971; Murdoch, 1973; Pulliam, 1974; Katz, 1974; Estabrook and Dunham, 1976; Covich, 1976). An optimal foraging strategy may be defined as that which yields the greatest net energy and nutrient gain per unit foraging time.

What may we expect of pollinators from foraging theory? Do pollinators behave in the anticipated manners? Answers to both questions will be briefly discussed. Expectations about behaviour are drawn from the predation literature and I will transpose them into a pollinator context. The papers cited do not discuss pollinators. Observations on pollinators of course are taken from the pollination biology literature.

Pollinators should differentiate between different plant species and form "search images" of the most favourable species. This behaviour promotes efficient localisation of time and effort and maximises benefit cost relationships (Tinbergen, 1960; Royama, 1970; Tullock, 1971). The tendency of individual pollinators to forage for a period of time within a plant species rather than foraging at random among several suitable species (the adaptation referred to as flower constancy) is a typical form of behaviour (Grant, 1949; Free, 1970; Baker & Hurd, 1968; Proctor & Yee, 1973; Heinrich, 1975).

Pollinator preference at some point in time is determined principally by the quantity and quality of the floral reward (Butler, 1945; Linsley, 1958; Free, 1968; Mosquin 1971; Heinrich 1975). The dominant factors affecting bee visits appear to be nectar abundance and sugar concentration and chemical attractants in pollen (Martin & McGregor, 1973). Hummingbird food selection is based primarily on sugar concentration and secondarily on rate of intake; sugar composition has little affect on choice (Hainsworth & Wolf, 1976). In a highland Costa Rican community, the hummingbird portion of the nectarivore guild seems to be organised around the foraging efficiency gradient of bird-plant interactions (Wolf, Stiles & Hainsworth, 1976). Within the quality gradient, each bird species first exploits plant species which it can forage most efficiently, then moves to the next level on the gradient.

Pollinators should become specialists when the density and quality of one or a few resource plants is high and if the resources are predictable (Levins & McArthur, 1969; Schoener, 1969; Colwell, 1973). Pollinators whose populations are food limited will seldom be food specialists (Emlen, 1973). Specialisation is a property of populations of pollinators rather than individuals, and under the aforementioned circumstances should enhance foraging efficiency. Specialisation and flower constancy are evolutionary and behavioural responses, respectively, to the absolute abundance of different food plants, not their relative abundance (Schoener, 1971; Pulliam, 1974; Estabrook & Dunham, 1976). Pollinator specialisation is most highly developed in the tropics (van der Pijl, 1969; Baker, 1970, 1973; Gentry, 1976), although it is observed to various degrees in temperate communities (Macior, 1971, 1974; Moldenke, 1975; Heinrich, 1976a). Specialization may reach its peak in the seasonal tropics where the flowering of different species is synchronised both within and between years (Janzen, 1967, 1974). A few specialised pollinators depend on one or two species during an entire season, but most switch from one diet to another in a regular way with the seasons. If the reward structure was constant in time, specialists would never have to leave the one to few species upon which they specialise. However, since resources change through time, even specialists must occasionally visit alternate species as a necessary compromise (Oster & Heinrich, 1976). This form of exploratory behaviour has been well-documented in individual bumblebees (Heinrich, 1976a).

Pollinators should avoid unfamiliar species even though they could be highly nutritious, palatable and conspicuous. This expectation is a correlate of flower constancy and specialisation (Manly, 1973). Avoidance of rare host appearances has been observed within species (Levin, 1972) and within species assemblages (Levin & Anderson, 1970; Macior, 1974).

Pollinators should minimise their foraging space (Covich, 1976). Site specificity reduces a forager's random movements, permitting it to maximise its energy return per energy investment. Site constancy is well documented in some bees, hummingbirds and butterflies, and may be in force for several hours to several days (Free, 1970; Levin & Kerster, 1974; Heinrich, 1975). Pollinators may spend hours foraging within an area of a few square metres and return to the same part of the population on a subsequent foraging bout. Even when artificially displaced, some pollinators will return to the initial capture site (Keller, Mattoni & Seiger, 1966). Foraging space is minimised by moving from a plant to one of its near neighbours, the exact distances covered being a function of plant density. As density declines, flight distance increases (Levin & Kerster, 1969; Wolf, 1969).

Another aspect of site specificity is the tendency of some pollinator species to repeat specific foraging paths, i.e., to trapline (Heinrich, 1975). These pollinators should be of relatively large size (Schoener, 1969; Colwell, 1973) since their resources are highly dispersed in space and time, although predictable in both dimensions. This pattern of foraging is found in some tropical bees (Janzen, 1971), bats (Heithaus, Opler & Baker, 1974; Heithaus, Fleming & Opler, 1975), hummingbirds (Colwell, 1973; Stiles, 1975) and butterflies (Gilbert, 1975). These pollinators are not loyal to a single species, but visit several with divergent floral adaptations, which permits some pollen to be transported between conspecific plants.

Pollinators should broaden their diet and switch hosts more frequently as their environment becomes more heterogeneous in space and time, and as patch size declines relative to foraging range (Schoener, 1974a; MacArthur, 1972; Emlen, 1973). Consider two habitats differing in patch size only; foraging time per item is the same. If a pollinator works only one species, travel time per item increases as patch size decreases because the distance between patches varies linearly with the linear dimension of a patch, whereas the foraging area within a patch varies as its square (Pianka, 1974). Therefore, as patch size decreases relative to the needs of a pollinator, patch selection becomes less advantageous. Switching hosts is an optimisation of two processes—search image formation and flexibility in plant species selection (Cornell, 1976). We observe that flower constancy and specificity decline as patch size declines (Free, 1970; Levin & Anderson, 1970; Heinrich & Raven, 1972; Heinrich, 1976a) and as the habitat becomes more unpredictable (Moldenke, 1975).

Pollinators should broaden their diets as the differences between patches decrease (Gillespie, 1974). Patches are recognised by the magnitude of the differences between them. The nutritional advantage of fidelity to one patch type declines as the utility of other interposed or juxtaposed patches increases, which in turn effectively reduces interpatch distances. We have little information on the reward structure of plants within the same or different patches. We can only suspect that pollinators are actually taking the magnitude of patch differences into account.

Pollinators should leave a patch when the rate of food intake in that patch drops to the average for the habitat as a whole (Charnov, Orians & Hyatt, 1976). The availability of pollen and nectar may be depressed as a consequence of foraging, thus lowering the subsequent profitability of foraging. Once resources have declined to some critical level or once the size of the pollinator population reaches some density threshhold, individuals should begin to emigrate to another patch of similar or different species composition (Parker & Stuart, 1976).

Emigration should commence when the future fitness due to emigration to another patch is greater than the expected future fitness due to continued occupation of a given patch (Gillespie, 1974). The expected effect of resource depletion, even though it may only be temporary, has been observed in pollinators of cultivated (Free, 1970) and wild plants (Heinrich, 1975; Wolf, Hainsworth & Gill, 1975). Limited rewards per unit time may be a device to force pollinators from one plant to another of the same species (Heinrich & Raven, 1972), but uniform depletion defeats this purpose.

Pollinators should contract their feeding habitats, but not their diet, when faced with competition from other vectors in ecological time. Diet should remain constant or expand (MacArthur & Wilson, 1967; Schoener, 1974a). However, if competition persists over long periods of time, and if resources are abundant, evolution may be expected to redistribute the phenotypes of the pollinator species and reduce the level of competition (Schoener, 1974b; Sale, 1974; Roughgarden, 1976). Competition for pollinator service and ostensibly among pollinators, has been described from numerous floras in the New World (Hocking, 1968; Mosquin, 1971; Kevan, 1972; Macior, 1973, 1975; Percival, 1974; Gentry, 1976; Stiles, 1975; Heinrich, 1975, 1976b). In a study of a tropical lowland community, Heithaus (1974) showed that pollinator niche breadth decreases as the diversity of pollinators increases. Resource partitioning by congeneric species has been demonstrated in bees (Johnson & Hubbell, 1975; Heinrich, 1976b; van der Pijl & Dodson, 1966), hummingbirds (Stiles, 1975; Gill & Wolf, 1975a, b), and butterflies (Gilbert & Singer, 1975; Schemske, 1976), and by the broad complement of pollinators exploiting Costa Rican species of *Cordia* (Opler, Baker & Frankie, 1975) and of several genera of the Bignoniaceae (Gentry, 1976).

Pollinators should expand their feeding habitats and diets when resources are limited and vacant niches are available, provided there is little competition (MacArthur, 1972; Keast, 1970; Yeaton & Cody, 1974; Grant, 1972). The sparser the resources, the less attention should be given to a patch, and the number of patches in a foraging bout should be increased (Schoener, 1969). Moreover, the less diverse the flowers and the reward of different plant species between patches, the less is the nutritional advantage of constancy or specialisation by pollinators, and the greater the potential advantage in foraging in a more promiscuous fashion. Pollinators are most likely to experience release from competition upon invading a distant island or a continental equivalent. Island floras are conspicuously depauperate in flora diversity, and for the most part are adapted to generalist pollinators

(Carlquist, 1974). Evidence on the foraging behaviour of island pollinators relative to their continental progenitors is meagre, but it appears that the former are more generalised feeders than the latter. Depauperate pollinator faunas, easy access to flowers and broad food niches on islands have been described by Carlquist (1974), Stern (1971), and Bowman (1966). The Afroalpine flora shows an interesting parallel with oceanic islands in the paucity of specialised pollinators and plants (Hedberg, 1957).

DEVELOPMENT AND REGULATION OF PATCH SIZE

Given that one or a few plant species form a dietary patch for a single pollinator species, from whence do patches arise? The spatial component of the environment operates to increase species diversity, and increase patch type but decrease patch size (S. Levin, 1974, 1976). This arises from the heterogeneity of the environment, especially with regard to edaphic and microclimatic differences, and interspecific competition. Patchiness also can arise in an initially homogeneous environment due to stochastic events such as colonisation pattern, whose effects may be magnified by species interactions. In essence, patchiness is self-augmenting (Whittaker, 1969). Patchiness of the environment also is promoted by disturbance, which provides the opportunity for local differentiation through random colonisation, and constantly interrupts the natural successional sequence on a local scale.

Operationally, the patch structure of an environment is that which is recognised by or relevant to a particular pollinator species (Wiens, 1976). Patchiness is thus pollinator-defined, and must be considered in terms of the perceptions of the pollinator rather than those of the investigator. The patch structure of a habitat thus will be interpreted variously by different species and types of pollinators. A patch will be a spatially discrete natural assemblage of suitable host plants which may be comprised of one to a few species at any point in time. A plant species could appear in more than one patch type. The assemblage of plants comprising a patch probably will be common ecological associates which frequently grow together within a given habitat.

If a pollinator perceives a patch as a population of one species, then there will be predictable changes in patch size during seral succession, and differences between species and communities whose populations have different factors regulating plant density. When a species enters a habitat, there will be a few initial nuclei which will be randomly distributed, and patch size will be small. As the species become established, the distribution of conspecifics will become more contagious or aggregated (and thus patch size will increase) due to the tendency of seedlings or vegetative offshoots to be narrowly dispersed and the coalescence of neighbouring patches. As the community matures, patch size declines with individuals achieving a more random distribution (Laessle, 1965; Margelef, 1958; Kershaw, 1958; Pielou, 1966; Brereton, 1971). Moreover each stage of succession displays a patch cycle like its constituents (Yarranton & Morrison, 1974). Therefore groups of species which may comprise the same patch may rise and fall in a roughly synchronous fashion.

Patchiness, in general, appears to decrease as communities mature along successional gradients, or as we go from regions of low to high abiotic and biotic environmental predictability with an attendant increase in species diversity and spatial complexity (Greig-Smith, 1961, 1964; D. Anderson, 1967; Whittaker, 1969; Williams, Lance, Webb & Dale, 1969; Goodall, 1970; MacArthur, 1972; Morrison & Yarranton, 1973; Kershaw, 1973). The change in patchiness as a function of maturity is due to an alteration in the factors which regulate population density. In immature communities, density regulation may be achieved principally by abiotic factors, and intra- and interspecific competition. In mature communities, regulation by herbivores and pathogens is paramount. Since the area over which these agents operate is much greater than that over which competition is effective, populations in mature communities, especially in some tropical regions, are thinned to a much greater degree than those in other areas (Janzen, 1970, 1972; Burdon & Chilvers, 1975; Cromartie, 1975; Tahvanainen & Root, 1972; Root, 1973; Strandberg, 1973).

Patches exist in time as well as space. Fugitive plants provide the most ephemeral patch type. In order of increasing patch persistence, there are the annuals and short-lived perennials of continuously disturbed communities, long-lived perennial herbs, and woody species. The latter are best represented in the non-seasonal tropics. There seems to be an inverse correlation between patch size and longevity of the constituents. Patches of fugitive species may become enormous in a few years, and go extinct prior to having an equilibrium density imposed upon them by herbivores and pathogens. Patches of tropical plants may be small due to devastating density-dependent control by pests, but may persist in equilibrium for many generations and hundreds of years.

THE GENETIC CONSEQUENCES OF FORAGING BEHAVIOUR

The foraging behaviour of pollinators determines their effectiveness as pollen vectors for single species and their penchant for interspecific pollination. How pollinators forage within a species will have a profound effect on the breeding structure and organisation of genetic variation within and among populations. Foraging in response to factors of distance and time will be a prime determinant of the quantity and distance of interspecific pollen hybridisation, cross-compatibility barriers permitting. Pollinator behaviour also may affect the nature and level of plant species diversity within communities. These issues have been discussed elsewhere (Levin & Anderson, 1970; Heinrich & Raven, 1972; Heithaus, 1974; King, Gallaher & Levin, 1975) and will not be discussed here. The models of foraging strategy on which the prescription for pollinator behaviour is based, although somewhat simplistic, seem to have good general predictive value for many foragers, especially pollinators.

Foraging patterns and interspecific hybridisation

Botanists have long had a love affair with interspecific hybridisation. There are hundreds of accounts of hybridisation, and considerable speculation on the factors influencing the ecogeographical distribution of the phenomenon. Hybridisation might be more common in temperate areas than in the tropics, in disturbed areas than those which are pristine, and on distant islands than on continents (Anderson, 1948, 1949; Anderson & Stebbins, 1954; Heiser, 1949, 1973; Carlquist, 1974). Presumably these patterns reflect, in part, the availability of suitable sites for the establishment of hybrids, and the potential value of introgression. Very little attention has been given to the fact that in zoophilous species, the incidence of hybridisation is a function of pollinator behaviour. Regardless of the habitat or the benefits of introgression, hybridisation is contingent upon interspecific pollination; this fundamental relationship must be included in explanations of hybridisation patterns.

Having considered foraging behaviour and patch size as dependent and independent variables, it is now possible to see what conditions and regions are most conducive to interspecific pollination, and thus to hybridisation. By definition, the frequency of hybridisation will be an inverse function of flower specialisation and constancy, since both constrain the wanderings of pollinators. As noted earlier, pollinator specialisation is an adaptation in evolutionary time, and constancy is an adaptation in ecological time; and they are not mutually exclusive. The general relationship between the two adaptations as they relate to patch switching and hybridisation is depicted in Fig. 1. We will assume that related species do not occur in the same patch. The three curves represent different levels of switching, and all points along a given curve represent the same level of hybridisation. Different combinations of specificity and constancy may provide the same result. A pollinator species will assume a position on the diagram as dictated by its long-term participation within a community (specificity) and its present perception of patch size and quality (constancy). If the level of specificity and constancy (in combination) are high, only a small proportion of all flights will be between patches and the level of hybridisation will be low. On the other hand, if generalised on both counts, the level of hybridisation will be relatively high. The position of a pollinator species is fixed on the Y-axis in evolutionary time (specialisation), but the position on the X-axis may vary as patch quality changes in time and space (constancy). The penchant for interspecific pollination will shift accordingly.

From foraging theory we may infer that pollinators are least likely to cross-pollinate two species when patch sizes are relatively large, and composed of high quality species which are persistent over a period of several years. Stringent interspecific competition among pollinators for limited resources will tend to reduce the number of patch types exploited, and thus is a deterrent to hybridisation. The potential for hybridisation also is low if the patches are very distinctive in composition. The foregoing suggest that patch-switching is likely to be relatively infrequent during a foraging bout in stabilised mid-successional and climax communities, geography notwithstanding. Since increasing plant species diversity is apparently accompanied by greater pollinator specialisation and smaller patch size, the position of a "typical" pollinator species should move along the outer curve from point A to point B in Fig. 1 as one moves from temperate to tropical communities.

The probability of interspecific pollination is greatest when patch size is relatively small and composed of resources which are unpredictable in time, and when neighbouring patches have similar nutritional value. Interspecific hybridisation is most likely to occur between fugitive or early to mid-successional species, especially during initial population growth when the congeners will be less aggregated into unispecific arrays. Hybridisation also has a relatively high probability of occurring in sites subject to periodic, but non-uniform disturbance, because this perturbation would diminish patch size and bring species with different ecological

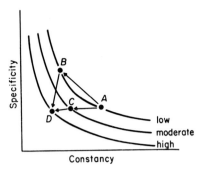

Figure 1. The relative levels of interspecific cross-pollination as a function of pollinator specificity and constancy. Lines of high, moderate and low levels of cross-pollination are shown. All points along a line represent the same level of cross-pollination.

requirements into closer proximity. Patch distintegration would move a pollinator species from points A to C, C to D, or A to D in Fig. 1, because flower constancy would decline. Hybridisation in late successional temperate genera is most likely to occur along suture zones formed as a consequence of post-glacial migration and range expansion (Remington, 1968). Large distant islands offer an excellent arena for interspecific pollen exchange and hybridisation because pollinators there tend to be generalists, and where patch size is small may display only moderate flower constancy. A species of continental pollinator located at point B in Fig. 1 might shift to point D were it transplanted to an island.

The proportion of pollinator flights between congeneric species in different patches is a crude indicator of the likely level of hybridisation. The cross-compatibility of species may vary from one habitat or community to another as may the extent of suitable sites for the establishment of hybrids. In spite of these constraints, pollinator foraging behaviour is a meaningful predictor of hybridisation levels. It is the first step in the chain of events necessary for the presence of hybrids.

The breeding structure of species

The breeding structure of single populations and groups of populations is poorly understood. Studies on crop plants and some native plants have demonstrated that populations are not panmicitic units, and that gene exchange between populations separated by a few hundred metres is rare, the vehicle of gene transport notwithstanding (Levin & Kerster, 1974). The breeding structure of species and single populations is determined by pollen dispersal, seed dispersal, and the incidence and nature of assortative pollination. Most assortative pollination is expressed as self-pollination and pollination between plants with close spatial proximity. Both forms of assortative pollination result in inbreeding; the first for obvious reasons, the second because the relationship between plants is typically an inverse function of distance. This relationship is due to restricted seed dispersal, most seeds being deposited in the vicinity of the seed parent (Levin & Kerster, 1974). For insect-pollinated plants then, the behaviour of the pollinator emerges as the prime determinant of the breeding structure of populations.

Optimal foraging theory proposes that where patch size is large, the flight distance between successively visited plants should be density-dependent, the greater the interplant spacing the greater should be the flight mean and variance. This relationship has been well documented in temperate bees and their major plants (Levin & Kerster, 1969; Estes & Thorp, 1975; Estes *et al.*, unpubl.; Schaal, unpubl.). Presumably the same holds for tropical bee-plant associations when patch size is large. A different form of foraging behaviour (referred to as trapline pollination) is displayed by certain tropical bees adapted to small patch size, i.e. widely dispersed conspecific plants (Janzen, 1971; Williams & Dodson, 1972; Frankie, 1976; Frankie, Opler & Bawa, 1976).

The distances pollinators fly between plants is not an accurate description of pollen dispersal distances, since some pollen may be carried past several plants before it is deposited on a stigma. The more flowers visited per plant, the smaller will be the pollen carry-over. However, if there is carry-over, its effect may not be substantial because flight direction seems to be near random with respect to the previous one, and resembles a drunkard's walk (Levin, Kerster & Niedzlek, 1971). With pollen carry-over, mean dispersal distance equals $X\Sigma\,(X\sqrt{n_i})$, where $X\,\Sigma$ is the flight mean, and $X\sqrt{n_i}$ the proportion of pollen deposited on the n_i plant (Levin & Kerster, 1969). Consider the effect of a liberal carry-over schedule. If 50% of a pollen collected on a plant is deposited on the next one, 25% on the second plant, 13% on the third, and 6% each on the fourth and fifth, the mean pollen dispersal distance would be only 30% greater than the mean flight distance.

Wright's (1940, 1943, 1951) neighbourhood, or isolation-by-distance, model has proved valuable in thinking about breeding structure. A neighbourhood is the area of a colony within which mating is assumed to be random. In his models, the reference colony for panmixis is one composed of dioecious individuals with equal numbers of both sexes, random mating, and a Poisson distribution of number of offspring per parent. A colony is said to have a genetically effective size, N_e, if it undergoes the same rate of decay of gene frequency variance as a reference colony of size N. The effective size of a neighbourhood is equivalent to the number of reproducing individuals with in a circle whose radius is equivalent to twice the standard deviation of the gene dispersal distance. A circle of this type will include 86.5% of the parents of the individuals at its centre. The individuals present may be used to estimate effective density if the colony is stable (in age structure) and stationary (in numbers) and if the distribution per parent of off-spring reaching maturity is Poisson (Kimura & Crow, 1963). Deviations from these conditions undoubtedly exist, and accordingly the effective density usually will be less than that observed (flowering plants) in the prescribed circle (Falconer, 1960). The neighbourhood area is N_e/d, where d is the genetically effective density. Genetically effective density is approximately the density of flowering plants.

Wright (1946) applies $N_e = 12.6\ \sigma^2 d$ as the neighbourhood size in a population of hermaphrodites where male and female gametes show the same amount of axial dispersion. Expanding Wright's equation so that gene dispersal is affected by pollen (p) and seeds (s) yields

$$N_e = 12.6d[(\sigma_p^2 + \sigma_s^2)/2] = 6.3d(\sigma_p^2 + \sigma_s^2).$$

Wright's equation is based on the assumption that populations do not move in space. Accordingly, there must be no net movement of pollen or seeds in a given direction. By substituting calculated variances that have zero means in this equation, we obtain

$$\sigma_p{}^2 = \Sigma(p_i - p)^2/N_p = \Sigma(p_i - 0)^2/N_p = \Sigma p_i{}^2/N,$$

$$\sigma_s{}^2 = \Sigma(s_i - s)^2/N_s = \Sigma(s_i - 0)^2/N_s = \Sigma s_i/N_s.$$

To bring pollen (haploid) and seed (diploid) dispersal into accord, we use one-half the absolute pollen dispersal. Combining the two dispersal components in the same equation gives

$$N_e = 6.3d(p_i{}^2/2N_p + s_i{}^2/N_s).$$

By incorporating the proportion of outcross progeny (r) into this equation, we arrive at:

$$N_e = 6.3dr(\Sigma p_i{}^2/2N_p + \Sigma s_i{}^2/N_s).$$

By employing the neighbourhood size as an indicator of breeding structure, we are relating gene flow to the decay of genetic variance. In amphimictic plants, the narrower the area from which parents are drawn and the stronger the correlation of parental genes by descent, the smaller is the neighbourhood size. Self-fertilisation greatly restricts neighbourhood size, because it represents zero gene dispersal by pollen.

Density-dependent pollen dispersal in bee-pollinated plants and zoophilous plants in general yield neighbourhood sizes which are roughly constant over a range of plant densities, and neighbourhood areas which increase as plant density declines (Levin & Kerster, 1974). These consequences of density-dependent pollen flow have some important genetic implications. Populations are buffered against the loss of genetic variability which would accompany pronounced downward fluctuations in population density or in the proportion of plants flowering in a given year. The neighbourhood size when considered over generations is the harmonic mean of the genetically effective densities of each of several generations, and harmonic means are dominated by low values (Wright, 1938). In order to see the buffering effect of density-dependent pollen flow, consider the consequences of fluctuations in population density over a twelve year period on neighbourhood size. Population density and the harmonic mean of neighbourhood size in each year for six hypothetical populations are presented in Fig. 2. The densities and pollinator response to these are based upon bee foraging behaviour on *Liatris* as described by Levin & Kerster (1969). The four densities considered are 1, 3, 5, and 11 plants/m², respectively. The variances of pollinator flight distance in the order of increasing plant density are 2.6, 1.4, 0.8, and 0.04 m², respectively. For comparative purposes, the effect of density fluctuations on a hypothetical wind-pollinated plant whose pollen dispersal is density-independent is also presented in Fig. 2. Anemophilous pollen dispersal varies from species to species. For small herbs, it may be similar to that generated by bees foraging in the dense *Liatris* population. Accordingly, that is used for illustrative purposes. The neighbourhood

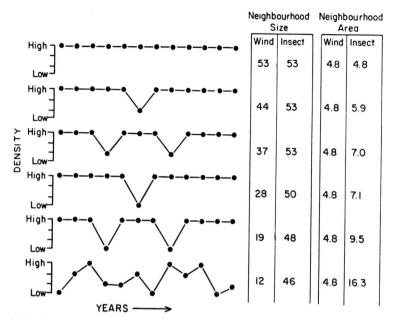

Figure 2. The effect of population size changes over 12 years on neighbourhood size and area, with density-dependent and density-independent pollination. Values are based upon pollen dispersal and densities described in the text.

size of populations of bee-pollinated plants is largely independent of density fluctuations of populations, whereas wind-pollinated plants experience declining neighbourhood sizes as populations decrease, and as mean densities decrease. The estimates of neighbourhood size are based only upon pollen dispersal.

Density-dependent pollen dispersal also affects neighbourhood area. As shown in Fig. 2, the mean neighbourhood area increases with declining density in bee-pollinated plants, but remains constant in wind-pollinated plants. Neighbourhood area is so defined that gene migration from a neighbourhood to an adjacent one occurs at a given rate per generation regardless of neighbourhood parameters. Therefore, the greater the number of neighbourhoods separating two points, the greater is the number of generations required for genes to flow from one point to the other. It follows that gene migration between two points will be much slower in high density arrays than in low density arrays, for neighbourhood diameters would be smaller in arrays where high density and close spacing prevailed.

Density and spacing will influence the ability of a population to undergo selective differentiation and subdivision through their effect upon gene migration. Should population dimensions remain constant, differentiation in response to a heterogeneous environment is most likely to occur in arrays of high density and close spacing, for under such circumstances gene migration is most restricted and swamping effected by migration is minimised. Similarly, high density arrays will be able to maintain many alleles at each locus at moderate frequency and will resemble a graded patchwork in general structure. In contrast, low density arrays will tend to display a more continuous variation pattern effected by a few "all-purpose" alleles at each locus.

The structure and gene flow potential within high density colonies bears a strong affinity to the population (= "colony system") structure which Wright deems most conducive to rapid evolution. Wright (1948) states that " . . . in a large population, divided and subdivided into partially isolated local races of small size, there is a continually shifting differentiation among the latter, intensified by local differences in selection, but occurring under uniform static conditions which inevitably brings about an indefinitely continuing, irreversible, adaptive, and much more rapid evolution of the species than in a comparatively large, random breeding population."

Levins (1964) proposed that gene flow among populations is part of the adaptive system of a species, and that optimum gene flow rates depend on the statistical structure of the environment. He argues that "the optimal amount of gene flow between populations is increased by the temporal variance of the environment variable . . .". The relationship between optimal gene flow and environmental stability permits the population to respond genetically to general, long-term environmental shifts while damping the response to local, short-term oscillations. On the basis of the arguments which were presented in this paper, we believe that Levins' rationale can be applied to gene flow within as well as between colonies. Furthermore, we have shown that density-dependent gene flow affords an ideal means of achieving a harmonious balance between flow rates and temporal environmental fluctuations.

The breeding structure of widely spaced plants or small clumps thereof is somewhat deceptive. Although plants are far apart, most pollination is between neighbouring plants, and seed dispersal is narrow. Thus widely spaced neighbouring plants in tropical forests may be sibs or half-sibs, as may be the case for closely spaced plants. If the feeding stations of trapline pollinators are small populations of conspecifics, the level of migration between neighbouring populations will be relatively high. Whether the feeding stations are single plants or small populations, the potential for local differentiation in the absence of strong selection will be small. The neighbourhood area may be thousands of square metres. The neighbourhood sizes probably would be large enough as to preclude random differentiation. Many tropical species whose plants are thinned to low densities have disproportionate sex ratios (Bawa, 1974; Bawa & Opler, 1975; Opler, Baker & Frankie, 1975; Gilbert, 1975, Opler & Bawa, 1977). A deviation from a 1 : 1 sex ratio reduces the effective density, which in turn reduces neighbourhood size, but increases neighbourhood area (Wright, 1931). Thus the potential for local differentiation in some tropical plants is further diminished. In these plants, the neighbourhood area must be tens of thousands of square metres. In summary, the effect of low density in some tropical forests is the same as it is elsewhere, except that greater interplant spaces can be tolerated because of the presence of trapline pollinators. In temperate areas, the paucity of conspecifics would not be possible, as they would tend to be ignored by pollinators, or receive insufficient conspecific pollen to insure adequate seed-set and plant replacement. Finally, it should be noted that such foraging behaviour on low density tropical plants is an evolutionary adaptation, and not an immediate behavioural response to density *per se*.

The discussion of density-dependent foraging has been focused upon bees, because their response to resource distribution and quality is far better understood than for the other groups of pollinators. However, the tendency of a pollinator to move from a plant to one of its near neighbours is evident in hummingbirds, bats, and butterflies and presumably occurs in moths, flies, and beetles as well (Levin & Kerster, 1974).

Consider next the breeding structure of a species whose populations typically are small (less than 30 plants) but relatively dense, and which are perceived of as a component of small patches by pollinators. Under these conditions, pollinators are expected to work most of the plants in the patch in a fashion more closely approximating to random mating than in large populations, and then move on to another nearby patch. If the same species is present in the other patch, inter-population pollen flow will ensue. Viewing populations as islands, this pattern of pollen and gene flow closely resembles the stepping stone model of Kimura & Weiss (1964). According to the stepping stone model, the gene pools comprising adjacent islands are correlated by descent to a greater degree than islands picked at random. Thus, the larger the number of islands lying between two islands of concern, the less correlated will be their gene pools, in spite of the fact that the absolute distance between the two islands remains the same. With small population sizes and considerable migration, the breeding structure of a species would encompass large numbers of plants and a large area. The opportunity for random differentiation between populations and establishment of regional variation patterns for characters free of selection is virtually nil (Kimura & Maruyama, 1971). The opportunity for selective differentiation between populations still exists, and most likely would be manifested in terms of clinal variation, rather than sharp discontinuities between neighbouring populations and local mosaic patterns of differentiation. The consequences of gene-flow selection balances between discrete populations in uniform and varying environments have been recently explored (May, Endler & McMurtrie, 1975; Nagylaki, 1976; Gillespie, 1976).

The foraging behaviour of pollinators has profound consequences on the breeding structure of populations and population systems and on the incidence of interspecific pollination, and thus has a profound effect on the amount and organisation of genetic variation within plants. Pollinators no longer should be viewed simply as agents of pollen transport, and populations no longer should be considered simply as cross-fertilising or partially self-fertilising. Studies of pollinator behaviour with reference to its genetic implications, and of population structure with reference to pollinators as causal agents are needed for a thorough understanding of plant population genetics within the context of the community.

ACKNOWLEDGEMENT

This study was supported in part by National Science Foundation grant DEB 76–19914.

REFERENCES

ALLARD, R. W., JAIN, S. K. & WORKMAN, P. L., 1968. The genetics of inbreeding species. *Advances in Genetics, 14:* 55–131.
ANDERSON, D. J., 1967. Studies on structure in plant communities, III. Data on pattern in colonizing species. *Journal of Ecology, 55:* 397–404.
ANDERSON, E., 1948. Hybridization of the habitat. *Evolution, 2:* 1–9.
ANDERSON, E., 1949. *Introgressive Hybridization.* New York: Wiley.

ANDERSON, E. & STEBBINS, G. L., 1954. Hybridization as an evolutionary stimulus. *Evolution, 8:* 378–389.

BAKER, H. G., 1970. Evolution in the tropics. *Biotropica, 2:* 101–110.

BAKER, H. G., 1973. Evolutionary relationships between flowering plants and animals in American and African tropical forests. In B. J. Meggers, E. S. Ayensu & W. D. Duckworth (Eds), *Tropical Forest Ecosystems in Africa and South America: a Comparative Review:* 145–160. Washington, D.C.: Smithsonian Institution.

BAKER, H. G. & HURD, P. D., 1968. Intrafloral ecology. *Annual Review of Entomology, 13:* 385–414.

BAWA, K. S., 1974. Breeding systems of tree species of a lowland tropical community. *Evolution, 28:* 85–92.

BAWA, K. A. & OPLER, P. A., 1975. Dioecism in tropical forest trees. *Evolution, 29:* 167–179.

BOWMAN, R. I., 1966. *The Galapagos.* Berkeley: Univ. of California Press.

BRERETON, A. J., 1971. The structure of the species populations in the initial stages of salt marsh succession. *Journal of Ecology, 59:* 321–339.

BURDON, J. J. & CHILVERS, G. A., 1975. Epidemiology of damping-off disease (*Pythium irregulare*) in relation to density of *Lepidium sativum* seedlings. *Annals of Applied Biology, 83:* 135–143.

BUTLER, C. G., 1945. The influence of various physical and biological factors of the environment on honey bee activity. An examination of the relationship between activity and nectar concentration and abundance. *Journal of Experimental Biology, 21:* 5–12.

CARLQUIST, S., 1974. *Island biology.* New York: Columbia University Press.

CHARNOV, E. L., ORIANS, G. H. & HYATT, K., 1976. Ecological implications of resource depression. *American Naturalist, 110:* 247–259.

COLWELL, R. K., 1973. Competition and coexistence in a simple tropical community. *American Naturalist, 107:* 737–760.

CORNELL, H., 1976. Search strategies and the adaptive significance of switching in some general predators. *American Naturalist, 110:* 317–320.

COVICH, A., 1974. Ecological economics of foraging among coevolving animals and plants. *Annals of Missouri Botanical Garden, 61:* 794–805.

COVICH, A., 1976. Analyzing shapes of foraging areas: some ecological and economic theories. *Annual Review of Ecology and Systematics, 7:* 235–257.

CROMARTIE, W. J., 1975. The effect of stand size and vegetational background on the colonization of cruciferous plants by herbivorous insects. *Journal of Applied Ecology, 12:* 517–533.

EMLEN, J. M., 1968. Optional choice in animals. *American Naturalist, 102:* 385–390.

EMLEN, J. M., 1973. *Ecology: an Evolutionary Approach.* Reading, Mass: Addison-Wesley.

ESTABROOK, G. A. & DUNHAM, A. E., 1976. Optimal diet as a function of absolute abundance, relative abundance, and relative value of available prey. *American Naturalist, 110:* 401–413.

ESTES, J. R. & THORP, R. W., 1975. Pollination ecology of *Pyrrhopappus carolinanus. American Journal of Botany, 62:* 148–159.

FAEGRI, K. & van der PIJL, L., 1971. *The Principles of Pollination Ecology,* 2nd ed. revised. Oxford: Pergamon.

FRANKIE, G. W., 1976. Pollination of widely dispersed trees by animals in Central America, with an emphasis on bee pollination systems. In J. Burley & B. T. Styles (Eds), *Tropical Trees. Variation, Breeding and Conservation:* 151–159. London: Academic Press.

FRANKIE, G. W., OPLER, P. A. & BAWA, K. S., 1976. Foraging behavior of solitary bees: implications for outcrossing of a neotropical forest tree species. *Journal of Ecology, 64:* 1049–1057.

FREE, J. B., 1968. Dandelion as a competitor to fruit trees for bee visits. *Journal of Applied Ecology, 5:* 169–178.

FREE, J. B., 1970. *Insect Pollination of Crops.* London: Academic Press.

GENTRY, A. H., 1976. Bignoniaceae of southern Central America: distribution and ecological specificity. *Biotropica, 8:* 117–131.

GILBERT, L. E., 1975. Ecological consequences of a coevolved mutualism between butterflies and plants. In L. B. Gilbert & P. H. Raven (Eds), *Coevolution of Animals and Plants:* 210–240. Austin: Texas. Univ. Texas Press.

GILBERT, L. E. & SINGER, M. C., 1975. Butterfly ecology. *Annual Review of Ecology and Systematics, 6:* 365–397.

GILL, F. B. & WOLF, L. L., 1975a. Foraging strategies and energetics of east African sunbirds at mistletoe flowers. *American Naturalist, 109:* 491–510.

GILL, F. B. & WOLF, L. L., 1975b. Economics of feeding territoriality in the golden-winged sunbird. *Ecology, 56:* 33–345.

GILLESPIE, J. H., 1974. Polymorphism in patchy environments. *American Naturalist, 108:* 145–151.

GILLESPIE, J. H., 1976. The role of migration in the genetic structure of populations in temporally and spatially varying environments: II. Island models. *Theoret. Population Biology, 10:* 227–238.

GOODALL, D. W., 1970. Statistical plant ecology. *Annual Review of Ecology and Systematics, 1:* 99–124.

GRANT, P. R., 1972. Convergent and divergent character displacement. *Biological Journal of the Linnean Society, 4:* 39–68.

GRANT, J., 1949. Pollination systems as isolating mechanisms in flowering plants. *Evolution, 3:* 82–97.

GREIG-SMITH, P., 1961. Data on pattern within plant communities. II. *Ammophila arenaria* (L.) Link., *Journal of Ecology, 49:* 703–748.

GREIG-SMITH, P., 1964. *Quantitative Plant Ecology*, 2nd ed. London: Butterworth.

HAINSWORTH, F. R. & WOLF, L. L., 1976. Nectar characteristics and food selection by hummingbirds. *Oecologia, 25:* 101–113.

HEDBERG, O., 1957. Afroalpine vascular plants. A taxonomic version. *Symb. bot. Upsal., 15:* 1–411.

HEINRICH, B., 1975. Energetics of pollination. *Annual Review of Ecology and Systematics, 6:* 139–170.

HEINRICH, B., 1976a. The foraging specializations of individual bumblebees. *Ecological Monographs, 46:* 105–128.

HEINRICH, B., 1976b. Resource partitioning among some unsocial insects: bumblebees. *Ecology, 57:* 874–889.

HEINRICH, B. & RAVEN, P. H., 1972. Energetics and pollination ecology. *Science, 176:* 597–602.

HEISER, C. B., 1949. Natural hybridization with reference to intergression. *Botanical Review, 15:* 645–687.

HEISER, C. B., 1973. Introgression re-examined. *Botanical Review, 39:* 347–366.

HEITHAUS, E. R., 1974. The role of plant-pollinator interactions in determining community structure *Annals of Missouri Botanical Garden, 61:* 675–691.

HEITHAUS, F. R., FLEMING, T. H. & OPLER, P. A., 1975. Foraging patterns and resource utilization in seven species of bats in a seasonal tropical forest. *Ecology, 56:* 841–854.

HEITHAUS, B. R., OPLER, P. A. & BAKER, H. G., 1974. Bat activity and pollination of *Bauhinia pauletia*: plant-pollinator co-evolution. *Ecology, 55:* 412–419.

HOCKING, B., 1968. Insect-flower associations in the high Arctic with special reference to nectar. *Oikos, 19:* 359–388.

JAIN, S. K., 1976. Evolution of inbreeding in plants. *Annual Review of Ecology and Systematics, 7:* 469–495.

JANZEN, D. H., 1967. Synchronization of sexual reproduction of trees within the dry season in Central America. *Evolution, 21:* 620–637.

JANZEN, D. H. 1970. Herbivores and the number of tree species in tropical forests. *American Naturalist, 104:* 501–528.

JANZEN, D. H., 1971. Euglossine bees as long-distance pollinators of tropical plants. *Science, 171:* 203–205.

JANZEN, D. H., 1972. Interfield and interplant spacing in tropical insect control. *Proceedings of the Annual Tall Timbers Conference, Ecology of Animal Control by Habitat Management:* 1–6.

JANZEN, D. H., 1974. Tropical black water rivers, animals, and insect fruiting by the Dipterocarpaceae. *Biotropica, 6:* 69–103.

JOHNSON, L. K. & HUBBELL, S. P., 1975. Contrasting foraging strategies and coexistence of two bee species on a single resource. *Ecology, 56:* 1398–1406.

KATZ, P. O., 1974. A long-term approach to foraging optimization. *American Naturalist, 108:* 758–782.

KEAST, A., 1970. Adaptive evolution and shifts in niche occupation in island birds. *Biotropica, 2:* 61–75.

KELLER, E. C., MATTONI, R. H. T. & SEIGER, M. S. B., 1966. Preferential return of artificially displaced butterflies. *Animal Behaviour, 14:* 197–200.

KERSHAW, K. A., 1958. An investigation of the structure of a grassland community. I. Pattern of *Agrostis tenuis. Journal of Ecology, 46:* 571–592.

KERSHAW, K. A., 1973. *Quantitative and Dynamic Plant Ecology*, 2nd ed. London: Edward Arnold.

KEVAN, P. G., 1972. Floral colors in the high Arctic with reference to insect-flower relations and pollination. *Canadian Journal of Botany, 50:* 2289–2316.

KIMURA, M. & CROW, J. F., 1963. The measurement of effective population number. *Evolution, 17:* 279–288.

KIMURA, M. & MARUYAMA, T. 1971. Pattern of neutral polymorphism in a geographically structured population. *Genetic Research, 18:* 125–131.

KIMURA, M. & WEISS, G., 1964. The stepping-stone model of population structure and the decrease of genetic correlation with distance. *Genetics, 49:* 561–576.

KING, C. E., GALLAHER, E. E. & LEVIN, D. A., 1975. Equilibrium diversity in plant-pollinator systems. *Journal of Theoretical Biology, 53:* 263–275.

LAESSLE, A. M., 1965. Spacing and competition in natural stands of sand pine. *Ecology, 46:* 65–72.

LEVIN, D. A., 1972. Low frequency disadvantage in the exploitation of pollinators by corolla variants in *Phlox. American Naturalist, 106:* 453–460.

LEVIN, D. A. & ANDERSON, W. W., 1970. Competition for pollinators between simultaneously flowering species. *American Naturalist, 104:* 455–467.

LEVIN, D. A. & KERSTER, H. W., 1969. The dependence of bee-mediated pollen and gene dispersal upon plant density. *Evolution, 23:* 560–571.

LEVIN, D. A. & KERSTER, H. W., 1974. Gene flow in seed plants. *Evolutionary Biology 7:* 139–220.

LEVIN, D. A., KERSTER, H. W. & NIEDZLEK, M., 1971. Pollinator flight directionality and its effect on pollen flow. *Evolution, 25:* 113–118.

LEVIN, S. A., 1974. Dispersion and population interactions. *American Naturalist, 168:* 207–228.

LEVIN, S. A., 1976. Spatial patterning and the structure of ecological communities. In S. A. Levin (Ed.), *Some Mathematical Questions in Biology, 7.* Providence, Rhode Island: Amer. Math. Soc.

LEVINS, R., 1964. The theory of fitness in a heterogeneous environment. IV. The adaptive significance of gene flow. *Evolution, 18:* 635–638.

LEVINS, R. & MacARTHUR, R., 1969. An hypothesis to explain the incidence of monophagy. *Ecology, 50:* 910–911.

LINSLEY, E. G., 1958. The ecology of solitary bees. *Hilgardia, 27:* 453–599.

LINSLEY, E. G., RICK, C. G. & STEPHENS, S. G., 1966. Observations on the floral relationships of the Galapagos carpenter bee. *Pan Pacific Entomology, 1:* 1–18.

MacARTHUR, R. H., 1972. *Geographical Ecology.* New York: Harper & Row.

MacARTHUR, R. & WILSON, E. O., 1967. *The Theory of Island Biogeography.* Princeton, N.J. Princeton Univ. Press.

MACIOR, L. W., 1971. Coevolution of plants and animals—systematic insights from plant-insect interactions. *Taxon, 20:* 17–28.

MACIOR, L. W., 1973. The pollination ecology of *Pedicularis* on Mount Rainier. *American Journal of Botany, 60:* 363–371.

MACIOR, L. W., 1974. Behavioral aspects of coadaptations between flowers and insect pollinators. *Annals of Missouri Botanical Garden, 61:* 760–769.

MACIOR, L. W., 1975. The pollination ecology of *Pedicularis* (Scrophulariaceae) in the Yukon Territory. *American Journal of Botany, 62:* 1065–1072.

MANLY, B. F. J., 1973. A linear model for frequency-dependent selection by predators. *Research in Population Ecology, 14:* 137–150.

MARGELEF, R., 1958. Information theory in ecology. *Genetis and Systematics 3:* 36–71.

MARTIN, E. C. & McGREGOR, S. E., 1973. Changing trends in insect pollination of commercial crops. *American Review of Entomology, 18:* 207–226.

MAY, R. M., ENDLER, J. A. & McMURTRIE, R. E., 1975. Gene frequency clines in the presence of selection opposed by gene flow. *American Naturalist, 109:* 659–676.

MOLDENKE, A. R., 1975. Niche specialization and species diversity along a California transect. *Oecologia, 21:* 219–242.

MORRISON, R. G. & YARRANTON, G. A., 1973. Diversity richness and evenness during a primary sand dune succession at Grand Bend. Ontario. *Canadian Journal of Botany, 51:* 2401–2411.

MOSQUIN, T., 1971. Competition for pollinators as a stimulus for the evolution of flowering time. *Oikos, 22:* 398–402.

MURDOCH, W. W., 1973. The functional response of predators. *Journal of Applied Ecology, 10:* 335–342.

NAGYLAKI, T., 1976. Dispersion-selection balance in localized plant populations. *Heredity, 37:* 59–67.

OPLER, P. A., BAKER, H. G. & FRANKIE, G. W., 1975. Reproductive biology of some Costa Rican Cordia species (Boraginaceae). *Biotropica, 7:* 234–247.

OPLER, P. A. & BAWA, K. S., 1977. Sex ratios of tropical dioecious trees. Selective pressures and ecological fitness. *Evolution:* in press.

OSTER, G. & HEINRICH, B., 1976. Why do bumblebees major? A mathematical model. *Ecological Monographs, 46:* 129–133.

PARKER, G. A. & STUART, R. A., 1976. Animal behavior as a strategy optimizer: evolution of threshold assessment strategies and optimal emigration thresholds. *American Naturalist, 110:* 1055–1076.

PERCIVAL, M. S., 1974. Floral ecology of coastal scrub in Southeast Jamaica. *Biotropica, 6:* 104–129.

PIANKA, E., 1974. *Evolutionary Ecology.* New York: Harper & Row.

PIELOU, E. L., 1966. Species-diversity and pattern diversity in the study of succession. *Journal of Theoretical Biology, 10:* 370–383.

van der PIJL, L., 1969. Evolutionary action of tropical animals on the reproduction of plants. *Biological Journal of Linnean Society, 1:* 85–92.

van der PIJL, L. & DODSON, C. H., 1966. Orchid flowers: their evolution and pollination. University of Miami Press, Coral Gables, Florida.

PROCTOR, M. C. F. & YEO, P. F., 1973. *The Pollination of Flowers.* London: Collins.

PULLIAM, H. R., 1974. On the theory of optimal diets. *American Naturalist, 108:* 59–74.

RAPPORT, D. J., 1971. An optimization model of food selection. *American Naturalist, 105:* 575–587.

REMINGTON, C. L., 1968. Suture-zones of hybrid interaction between recently joined biotas. *Evolutionary Biologist, 2:* 321–428.

ROOT, R. B., 1973. Organization of a plant-arthropod association in simple and diverse habitats: the fauna of collards. *Ecological Monographs, 43:* 95–124.

ROUGHGARDEN, J., 1976. Resource partitioning among competing species—a coevolutionary approach. *Theoretical Population Biology, 9:* 388–424.

ROYAMA, T., 1970. Factors governing the hunting behaviors and selection of food by the Great Tit (*Parus major* L.). *Journal of Animal Ecology, 39:* 619–668.

SALE, P. F., 1974. Overlap in resource use and interspecific competition. *Oecologia, 17:* 245–256.

SCHEMSKE, D. W., 1976. Pollinator specificity in *Lantana camara* and *L. trifolia* (Verbenaceae). *Biotropica, 8:* 260–264.

SCHOENER, T. W., 1969. Optimal size and specialization in constant and fluctuating environments: an energy-time approach. *Brookhaven Symposium on Biology, 22:* 103–114.

SCHOENER, T. W., 1971. Theory of feeding strategies. *Annual Review of Ecology and Systematics, 2:* 369–404.

SCHOENER, T. W., 1974a. The compression hypothesis and temporal resource partitioning. *Proceedings of the National Academy of Sciences, 71:* 4169–4172.

SCHOENER, T. W., 1974b. Resource partitioning in ecological communities. *Science, 185:* 27–39.

STERN, W. L., 1971. *Adaptive Aspects of Insular Evolution.* Pullman: Washington State Univ. Press.

STILES, F. G., 1975. Ecology, flowering phenology and hummingbird pollination of some Costa Rican *Heliconia. Ecology, 56:* 285–301.

STRANDBERG, J., 1973. Spatial distribution of cabbage black rot and the estimation of diseased plant populations. *Phytopathology, 63:* 998–1002.

TAHVANAINEN, J. O. & ROOT, R. B., 1972. The influence of vegetational diversity on the population ecology of a specialized herbivore *Phyllotreta* (Coleoptera: Chrysomelidae). *Oecologia, 10:* 321–346.

TINBERGEN, L., 1960. The natural control of insects in pine-woods. I. Factors influencing the intensity of predation by songbirds. *Archives néerlandaises de zoologie, 13:* 265–343.

TULLOCK, G., 1971. The coal tit as a careful shopper. *American Naturalist, 105:* 77–80.

WHITTAKER, R. H., 1969. Evolution of diversity in plant communities. *Brookhaven Symposium on Biology, 22:* 178–196.

WIENS, J. A., 1976. Population responses to patchy environments. *Annual Review on Ecology and Systematics, 7:* 81–120.

WILLIAMS, N. H., & DODSON, E. H., 1972. Selective attraction of male euglossid bees to orchid floral fragrances and its importance in long distance pollen flow. *Evolution, 26:* 84–95.

WILLIAMS, W. T., LANCE, G. N., WEBB, L. J. & DALE, M. B. 1969. Studies in the numerical analysis of complex rain-forest communities. III. Analysis of successional data. *Journal of Ecology, 57:* 515–535.

WOLF, L. L., 1969. Female territoriality in a tropical hummingbird. *Auk., 85:* 490–504.

WOLF, L. L., HAINSWORTH, F. R. & GILL, F. B., 1975. Foraging efficiencies and time budgets in nectar-feeding birds. *Ecology, 56:* 117–128.

WOLF, L. L., STILES, F. G. & HAINSWORTH, F. R., 1976. Ecological organizations of a tropical highland hummingbird community. *Journal of Animal Ecology, 45:* 349–379.

WRIGHT, S., 1931. Evolution in mendelian populations. *Genetics, 16:* 97–159.

WRIGHT, S., 1938. Size of population and breeding structure in relation to evolution. *Science, 87:* 430–431.

WRIGHT, S., 1940. Breeding structure of populations in relation to speciation. *American Naturalist, 74:* 232–248.

WRIGHT, S., 1943. Isolation by distance. *Genetics, 28:* 114–138.

WRIGHT, S., 1946. Isolation by distance under diverse systems of matings. *Genetics, 31:* 39–59.

WRIGHT, S., 1948. On the roles of directed and random changes in gene frequency in the genetics of populations. *Evolution, 2:* 279–295.

WRIGHT, S., 1951. The genetical structure of populations. *Annals of Eugenics, 15:* 323–354.

YARRANTON, G. A. & MORRISON, R. G. 1974. Spatial dynamics of a primary succession: nucleation. *Journal of Ecology, 62:* 417–428.

YEATON, R. I. & CODY, M. L., 1974. Competitive release in island song sparrow populations. *Theoretical Population Biology, 5:* 42–58.

Plant-animal interactions affecting gene flow in *Viola*

A. BEATTIE

Department of Biological Sciences, Northwestern University, Evanston, Illinois, U.S.A.

Most species of *Viola* exhibit a diplochorous seed dispersal system in which ballistically scattered seeds are located by ants and removed to nests. Ant activity transforms a random distribution to a highly clumped one which is reflected in high coefficients of dispersion for seedlings. The initial ballistic phase yields the greatest axial dispersal distances and it may be, in addition, a predator avoidance mechanism. The second phase of ant transportation generally involves shorter distances, since nests are numerous. Nests and their immediate environs are considered to be 'safe sites' for germination and establishment. Seeds scarified by ants germinate faster than those which never interact with ants.

The clumping of plants generated by ant activity is reinforced by pollinator service, involving principally solitary bees. Pollinator flight distances exhibit leptokurtic distributions. Among violet colonies with low spacing means and variances, flight distances are short, interplant flight frequencies are high and greater quantities of pollen are moved. On the other hand, colonies with high spacing parameters show longer pollinator flights but reduced interplant pollen exchange.

The activities of pollen and seed vectors tend to establish neighbourhoods of small size and area; levels of within-neighbourhood gene exchange are greater than levels of between-neighbourhood gene exchange. Subdivision of populations and differentiation among breeding units is to be expected under this regime. The regulation of gene flow by insect mutualists is likely to contribute powerfully to the complex patterns of variation seen in many species and groups of species in the genus.

KEY WORDS:—*Viola*—Virginia—ants—seed-dispersal—ballistic—pollination—safe-sites—cleistogamy—neighbourhood

CONTENTS

INTRODUCTION

A wide variety of animal species, both invertebrate and vertebrate, interact with the flowers, fruits and seeds of *Viola* populations. They may be conveniently divided into four categories for the purposes of the following discussions. The first two, pollen vectors and seed vectors, are involved in mutualistic interactions gaining

151

food from the plants which in turn benefit from the dispersal of their pollen and seeds. The second two, flower predators and seed predators, are entirely destructive to the plants, and I shall present evidence that they have been important selective agents in the evolution of *Viola*. The purpose of this paper is to suggest ways in which these interactions affect the dispersion of plants, the structure of populations and the potential for gene flow.

Some of the material to be presented comes from work done in England, California and Colorado. However, most data result from work carried out in the Monongahela National Forest in West Virginia where several study sites are located. These are in areas of 50 to 100 year-old secondary tree growth characteristic of both central and northern hardwood forests (Strausbaugh & Core, 1970). Thus, species such as *Acer saccharum, Fagus grandifolia, Pinus strobus, Carya ovata, Liriodendron tulipifera* and *Carpinus caroliniana* are common. The herbaceous flora of the forest floor is very rich and this is reflected in the fact that 16 species of *Viola* can be found within walking distance of our base camp. Nine species have been studied, seven from the section Nomimium, *V. cucullata, V. papilionacea, V. triloba, V. pedata, V. blanda, V. rostrata* and *V. striata*. The first five are stemless with pedicels arising directly from the rootstock. The last two are caulescent, the pedicels arising from the axils of leafy shoots. *V. blanda* differs from all the other species in being capable of vegetative reproduction. However, it has white flowers in common with *V. striata*, while the remaining species have blue, lilac or violet-purple flowers. The section Chamaemelanium is represented by *V. pensylvanica*, a caulescent species with yellow flowers. The predominantly Eurasian section which includes all the pansies—Melanium—is represented by the unique annual, north American species *V. rafinesquii* which is caulescent with small, variably coloured flowers.

SEED VECTORS

Seed vectors influence violet populations in at least three ways: (a) they disperse seeds, (b) they affect germination rates and (c) they are a major factor in determining the distribution (patterning) of adult plants. Seed dispersal takes place in two stages in most species of *Viola*. When a capsule is mature it is held away from the parental foliage on a stiff, erect peduncle. As the individual valves of the capsule dry out, the seeds are ejected and scattered around the parent plant, sometimes to distances of several metres (Fig. 1, Table 1). The second stage of dispersal is affected by ants. The seeds each bear a small external food body or elaiosome which is attractive to ants for several days. Consequently, the majority of them are located and removed by ants within hours of dehiscence (Beattie & Lyons, 1975). Careful observation has shown that most seeds are carried directly back to nests (Culver & Beattie, 1978) (Table 2). Species of the genera *Lasius, Aphaenogaster, Myrmica* and *Formica* are the ants principally involved in these interactions in West Virginia.

During the summer of 1976, seventy-five seed-choice or 'cafeteria' experiments were performed to discover whether or not there were preferences among ant species for particular violet species. In each experiment four fresh seeds were placed in a 1 cm square on the ground. The seeds came from violet species in the immediate vicinity and the experiments were conducted during the actual period of dehiscence of these plants. Preferences were identified by scoring for both the frequency and

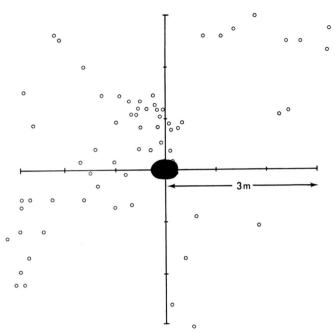

Figure 1. Example of pattern of seeds dispersed ballistically from a single adult *Viola papilionacea* in the greenhouse.

Table 1. Comparison of ballistic and ant dispersal distances of chasmogamous seeds. Ballistic data were taken in a greenhouse (see Beattie & Lyons, 1975). Student's *t*-test (two-tailed) was used to compare dispersal distances

Species		Ballistic dispersal				Ant dispersal to nest			
	n	X (cm)	Max.	S.E.	*n*	X (cm)	Max.	S.E.	*P*
V. rostrata	105	120	420	8	13	72	125	7	<0.05
V. pensylvanica	271	120	540	10	12	72	108	10	>0.10
V. pedata	250	140	510	5	30	35	150	8	<0.005

intensity of interactions. Intensity was indicated by five categories of behaviour: (1) ignore, (2) antennate, (3) examine, (4) attempt to pick up seed, (5) remove. While most seeds were removed the analysis of the data of the removal category showed no clear preferences on the part of any particular ant species for any particular violet species, for example see Table 3.

It is also possible for seed vectors to influence germination. To test this, colonies of ant species from West Virginia known to carry *Viola* seeds were brought into the laboratory. The colonies were housed in aluminium covered test tubes placed in plastic boxes where they thrived for many months. In a series of trials, seeds of *V. papilionacea* were divided into two lots. Those from the first were stored for several months at 5°C and then placed near to ant colonies and watched. The second lot remained in storage. All the seeds placed near to the colonies were quickly removed to the nests. Within two days most had been returned to the nest entrance intact, but with the elaiosome removed and the testa scarified. Next, both

Table 2. *Proportions of seeds transferred from the forest floor to ant nests for four species of* Viola *in West Virginia*

Name of study site	Violet species	No. of seeds	% taken to nest
Powerline	*V. pedata*	35	86
Powerline	*V. triloba*	26	100
Big Draft	*V. pensylvanica*	14	86
Anthony Creek	*V. rostrata*	15	87

lots of seeds were tested for germination under identical conditions. Seeds were placed in petri dishes on moist filter paper under the following temperature and light regime: 12 hours darkness at 8°C and 12 hours of light at 15°C. When seedlings reached a standard size they were planted in individual pots. The germination rate was greater among the ant manipulated seeds (64% compared to 56% for those which had remained in storage). After 45 days 33% of the ant-manipulated seeds had become juveniles with at least one adult leaf while only 12% of the control group had reached a similar stage. This difference was not quite significant ($P > 0.05 < 0.10$, Fisher's Exact Test). However, the data support a previous suggestion by Dymes (1916) that ant-manipulated seeds germinate faster and give rise to more vigorous plants with bigger root systems. It is possible that this confers a selective advantage through increasing the probability of establishment and a shortening of the period to first reproduction (Rabotnov, 1969).

In the same paper Dymes also suggested that seeds would be unlikely to grow in a nest which was still active. In a study of nest movement among the West Virginian ant species which carried *Viola* seeds, it was found that the average residence time at a particular site was only 30 days (Culver & Beattie, 1978). This would be of significance for two reasons: firstly, if seeds and seedlings are subject to mortality as a result of continuing ant activity at the nest, perhaps excavating around roots or gnawing shoots, they do not have to wait long before they may develop undisturbed. Secondly, if residence time is short, the number of seeds brought back to a nest may be so small as to minimise the potential problem of seedling competition at the nest site.

Table 3. *Analysis of some seed removal data to detect possible ant species preferences for* Viola *species, from two sites in West Virginia*

	Ants	Seeds		
		V. triloba	*V. pedata*	
Site 1	*Aphaenogaster* spp.	12	5	
	Myrmica punctiventris	9	1	
		$x^2_{(1)}$ — 1.33 N.S.		
		V. pensylvanica	*V. rostrata*	*V. papilionacea*
Site 2	*Aphaenogaster* spp.	16	7	9
	Lasius alienus	2	4	1
		$x^2_{(2)}$ — 3.71 N.S.		

Table 4. Coefficients of dispersion (variance: mean ratio giving a measure of randomness or non-randomness) for seedlings and adults of four species of *Viola* in West Virginia. Each value is calculated from a minimum of twenty 1-m² quadrats. All values subjected to a *t*-test and showed $P < 0.001$ for non-randomness (clumping)

	Seedlings	Adults
V. pedata	8.05	16.54
V. blanda	—	9.13
V. rostrata	7.42	7.92
V. pensylvanica	4.33	6.42

If establishment is most frequent in nest sites, the dispersion of plants, especially seedlings, would be non-random and highly clumped. This is in fact the case as can be seen in Table 4. An analysis of the dispersion of 30 forest floor herb species (Struik & Curtis, 1962) has shown that those showing such a strong degree of aggregation comprised a small minority. Inspection of the data shows that about half of them were known to be ant dispersed (Handel, 1976). Although the seed crop of an individual parent plant has not been followed through both the ballistic and the myrmecochorous stages of dispersal, the data in Table 4 strongly suggest that ant activity transforms a broad scatter of ballistically dispersed seeds (see Fig. 1) into a small number of clumps of seedlings.

SEED PREDATORS

The question now arises as to the specific advantages of the ballistic stage of dispersal. Ballistics achieve dispersal distance, but co-evolution with larger, more widely foraging ant species probably could have produced the same distribution of dispersal distances. Janzen (1970) introduced the idea that seeds clustered around the parent plant are particularly vulnerable to predators. While the idea was originally applied to tropical trees, Platt (1976) has shown that it is true for a perennial herb, *Mirabilis hirsuta*. In this case mice eat most seed within 50 cm of the parent but survivorship of seeds increases rapidly after this point. *Viola* has a number of important seed predators, especially small mammals (Ridley, 1930; Martin, Zim & Nelson, 1951; Gorecki & Gebczynska, 1962; Drozdz, 1966), but also birds (Ridley, 1930; Martin *et al.*, 1951; Witherby, Jourdain, Ticehurst & Tucker, 1952). Deer mice (*Peromyscus* spp.) are very common in West Virginian

Table 5. Rates of seed removal in exclosures (with ant removal) compared to non-exclosed seeds in West Virginia. Only *Viola papilionacea* seeds (of cleistogamous origin) were used

	Number of days		
	0	3	7
Seeds remaining: exclosure	60	45	38
Seeds remaining: non-exclosed	30	5	0
Difference in frequency taken	0	0.58	0.63
P (Fisher's Exact Test)		<0.001	<0.001

Table 6. Summary of regression data for plant spacing and pollinator flight parameters

Plant parameter	Flight parameter	P	Solitary bees				P	Lepidoptera			
			Slope	S.E.	Intercept	S.E.		Slope	S.E.	Intercept	S.E.
Spacing mean	Flight mean	0.001	0.346	0.06	0.067	0.20	—	0.520	0.37	1.528	0.59
Spacing variance	Flight variance	0.001	0.630	0.14	−0.005	1.26	0.05	3.374	1.29	9.452	6.28

forests and trials with wild-caught animals in the laboratory show that they consume violet seeds very readily. The seeds are totally destroyed by the action of the jaws and gut. Last summer a series of experiments were performed in West Virginia to establish the predation rates on violet seeds (Culver & Beattie, 1978). Exclosures which kept out small mammals and birds but admitted ants were placed over square grids of ten seeds placed at 5 cm intervals. Control grids were not exclosed. Assuming equal rates of ant removal of seeds in both experiment and control, almost 60% of the seeds were taken by predators in three days (Table 5).

Since this data suggests that seeds ejected singly from parent plants may still suffer high mortality from predators, it follows that accumulations of seeds immediately beneath adult plants might result in very great losses of seed crops. The ballistic stage of dispersal would minimise seed predation by reducing the probability of vulnerable aggregations of seeds. Thus, diplochory in *Viola* may involve the avoidance of one interaction, predation, without compromising another, i.e. mutualism with seed vectors.

POLLEN VECTORS

Most species of *Viola* are visited by a variety of insects including bumblebees (*Bombus*), solitary bees (e.g. *Osmia*, *Andrena*, *Halictus*, *Lasioglossum*), syrphid flies (e.g. *Rhingia*), beeflies (e.g. *Bombylius*), butterflies (e.g. *Pieris*) and moths (e.g.

Table 7. Summary of regression data for plant spacing parameters and percentage interplant flights for few-flowered, many-flowered and all species together. (Values calculated from arcsine transformations of interplant flight data)

Plant parameter	Flight parameter	P	Solitary bees				P	All visitors			
			Slope	S.E.	Intercept	S.E.		Slope	S.E.	Intercept	S.E.
Spacing mean	Percent interplant flights										
Few-flowered		0.05	−2.038	0.88	99.899	6.50	—	−1.689	0.69	85.136	4.21
Many-flowered		0.001	−1.449	0.33	73.770	6.05	—	−2.365	0.82	82.106	5.70
All species		0.02	−1.322	0.11	86.783	5.02	0.05	−2.579	0.40	85.860	3.76
Spacing variance	Percent interplant flights										
Few-flowered		0.05	−1.454	0.56	95.32	4.67	—	−1.010	0.51	85.040	3.99
Many-flowered		0.001	−3.399	0.46	85.996	9.28	—	−1.201	0.69	78.916	5.99
All species		0.01	−2.574	0.69	92.288	5.72	0.05	−1.138	0.44	82.436	3.51

Hemaris) (Beattie, 1971, 1972, 1974). Most of these insects are capable of pollinating violet flowers (Beattie, 1971, and unpubl. data) but for many species, especially
in eastern North America, solitary bees are the most constant and systematic
pollinators.

In West Virginia, pollination was analysed in terms of: (a) the frequency distribution of pollinator flight distances; (b) the frequency of within- and between-plant
pollen transfer; and (c) the actual numbers of pollen grains being moved. Items (a)
and (b) were obtained by careful observation of pollinator behaviour in natural
populations, and item (c) was obtained by examining the pollen content of stigmatic
cavities (Beattie, 1969) from post-anthesis flowers in the same populations. I shall
discuss each item in turn.

Table 8. Data on the frequency distributions of pollen dispersal (insect flight)
distances and seed dispersal (ballistic) distances for four species of *Viola* in West
Virginia. $t_{(1)}$ = t-test for null hypothesis of equal means, $t_{(2)}$, $t_{(3)}$ = t-test for null
hypothesis of zero kurtosis and zero skewness respectively

	Number of observations	Mean	$t_{(1)}$	Variance	Kurtosis	$t_{(2)}$	Skewness	$t_{(3)}$
Viola pensylvanica								
pollen	75	71.9		4922.2	3.61	0.001	1.90	0.001
			0.001					
seeds	268	114.9		7155.8	2.59	0.001	1.50	0.001
Viola rostrata								
pollen	92	14.3		2929.0	5.62	0.001	2.37	0.001
			0.001					
seeds	109	115.2		6992.2	2.56	0.001	1.58	0.001
Viola blanda								
pollen	111	37.0		2111.5	21.1	0.001	3.88	0.001
			0.001					
seeds	17	79.8		2854.3	−1.73	N.S.	−0.36	N.S.
Viola pedata								
pollen	230	49.3		5027.0	6.76	0.001	2.57	0.001
			0.001					
seeds	232	128.8		4136.0	1.05	0.01	−0.89	0.001

The frequency distributions of pollinator flight distances for almost all species
showed marked leptokurtosis (Table 8). In other words, the insect pollinators were
mostly travelling only short distances (mean for all species = 50.6 cm), while a small
fraction travelled longer distances, occasionally up to 30 m. This was reflected in the
tendency toward positive skewness in the same frequency distributions (Table 8). A
second important finding was that pollinator flight parameters (mean, standard
deviation and variance) were positively correlated with plant spacing parameters
(mean, standard deviation and variance of distances between plants) (Table 6).
These correlations held over a wide range of plant densities. Similar correlations
have been discovered before (Levin & Kerster, 1969), and they suggest that pollen
dispersal is directly related to the spacing of pollinator resource plants on the
ground.

Table 7 summarises correlations between plant spacing parameters and the percentages of pollinator flights which were between plants. The data are divided according to different flowering regimes; (a) species with one or two flowers open simultaneously; (b) species with more than two flowers open simultaneously; and (c) all species lumped together. The proportion of between-plant flights was always greater than 50%, regardless of density. Significant negative correlations showed that as spacing parameters (mean and variance) increased, the frequency of between-plant flights decreased. This means, for example, that when spacing parameters are high (and plant density low) the pollinators move between plants less frequently and pollen dispersal is likely to diminish.

Finally, a significant negative correlation was obtained for the relationship between plant spacing means and the numbers of pollen grains being moved by pollinators (Beattie, 1976). Thus, continuing from the example just given, when the spacing means are high or variable, fewer pollen grains are being exchanged between plants.

FLOWER PREDATORS

This category includes nectar thieves such as bibionid, muscid and calliphorid flies (Beattie, 1972) and bumblebees which learn to bite through the nectar spur of the corolla, thus failing to operate the pollination mechanism (Beattie, 1971).

The flowers and ovules may also be damaged, and frequently totally destroyed by lepidopteran larvae (Ehrlich & Ehrlich, 1961; Higgins & Riley, 1970), cecidomyiid larvae (D. Leatherdale, pers. comm.) and various unidentified dipteran larvae, snails and slugs (Beattie, unpubl. data). This topic has not been studied systematically but the occurrence of various potential toxins such as alkaloids, glycosides and flavonoids in violet tissues (Frenclowa, 1962; Paris, 1963; Horhammer et al., 1965) suggest that predispersal predation of seeds, particularly at the ovule stage, may be an important selective pressure.

CHARACTERISTICS OF THE MUTUALISMS

Competition and predation are the best understood species interactions while mutualisms are scarcely understood at all. Plant-pollinator and plant-ant interactions are cited as examples of mutualisms but the sense of much of the literature is that they are exotic products of evolution, generally occurring in the tropics or subtropics, almost entirely obligate and involving either a pair of taxa or a very small group of them (Emlen, 1973; Pianka, 1974; Ricklefs, 1973). By contrast the work with *Viola* emphasises that mutualisms are probably very common evolutionary processes which are frequent in the temperate zones and which can involve complex assemblages of species which show a range of responses, from facultative to obligate.

The data on violet pollen and seed vectors shows that the degree of interdependence is very variable and that the effects can be asymmetrical. Seed set in some violet species depends entirely upon insect pollinators which also forage among the flowers of a wide variety of other plant genera. In this case the mutualism is obligate for the plants but not for the pollinators. An example is *Viola pedata* which has no cleistogamous flowers. In this species the chasmogamous flowers are very large and colourful and attract an unusual diversity and abundance of potential pollinators (Beattie, unpubl. data). Other species, by contrast, set most seed cleistogamously, produce small flowers and attract relatively few insect visitors (e.g. Clausen, Channel & Uzi Nur, 1964). Here, the degree of inter-dependence appears to be very low. For some pollinator species which emerge in early spring when flowers are scarce, *Viola* can be a vital resource (Beattie, 1969). Many early-flowering violets have alternative means of reproducing besides chasmogamy and so the dependence is largely on the part of the pollinator. How-ever, this dependence may diminish as the season progresses and alternative resource plants come into flower, so that even such an important pollinator as *Andrena violae*, reputedly an obligate mutualist, can be seen to be feeding on many other plants besides violets in West Virginia.

With respect to seed vectors, the data indicate that no violet species is dependent on one particular ant species, or even a closely related group of species (Table 3). Also, no ant species appears to require violets for survival. In early summer many ant species take seeds eagerly and quickly but this enthusiasm tapers off towards late summer.

The mutualisms of which *Viola* populations comprise a part, show great vari-ation in the degrees of interdependence and in the assemblages of species involved. It is of interest that the vectors required by the plants are not characterised by any particular taxonomic affinity but rather by the possession of particular morpholo-gies and behaviour patterns. Most species of *Viola* rely on visits from a broad cross-section of small bodied or long-tongued spring-flying insects for cross-pollination. Similarly, most also are aided by the presence of some medium-sized, ground foraging ant species for seed dispersal. These mutualisms may be the most depend-able, particularly in temperate zones where weather patterns render the distribution and abundance of vectors less predictable. The point is that complex mutualisms are buffered against spatial and temporal discontinuities in the availability of any given vector; pollination and seed dispersal may continue more or less un-interrupted. In this way the potentially critical functions of reproduction by seed, and gene flow, are maintained to some degree.

Finally, the interactions between the predators and the mutualists are almost wholly unexplored. The predators and the mutualists may themselves be viewed as competitors for the floral rewards or the seeds. Graphical models of pairwise interactions show that as the interdependence of a mutualism increases, its stability decreases, so that populations tend either towards extinction or toward indefinite growth (D. Culver, pers. comm.). The interaction may be stable, how-ever, if there is an upper limit to the population sizes of both species. In the *Viola* system, the predators may be part of a complex of factors which limit population increase among the mutualists. The unexpected result from this form of model is that predators may increase the stability of the mutualisms responsible for pollen and seed dispersal (Ray Heithaus, pers. comm.).

PLANT PATTERNING

Another factor possibly limiting population sizes of the mutualists, at least the plants, is the availability of safe sites (Harper, Clatworthy, McNaughton & Sagar, 1961). The data in Table 1 show that mean ballistic dispersal distances are greater than mean ant dispersal distances. In addition seeds may be carried by ants either towards or away from parent plants. Consequently, the major selective advantage of ant transport is more likely to be relocation to a safe site for germination and establishment than increase in dispersal distance. The potential importance of seeds being 'planted' by ants should not be underestimated. Many authors have emphasised that the plant environment is extremely heterogeneous or patchy with respect to suitable sites for germination and establishment (S. A. Levin, 1976; Wiens, 1976). If ant nests or their immediate environs are prime safe sites, then among subcommunities or guilds of myrmecochores, the abundance and distribution of safe sites may be important determinants of plant dispersion, population structure and hence gene flow.

Violet seeds scattered across the forest floor by ballistic mechanisms alone encounter an environment which is highly heterogeneous with respect to three crucial factors: (1) predation, (2) stimuli necessary for germination and (3) nutrients necessary for establishment and growth (Young, 1934; Stickel & Warbach, 1960; Frankland, Ovington & Macrea, 1963; Harper, Williams & Sagar, 1965; Riese & Wiedemann, 1975; Bratton, 1976). Nutrient levels, or example, can change markedly over distances as short as 10 cm in an area of forest floor which, to the eye, appears very uniform.

In West Virginia, the violet species under discussion grow on grey-brown podzols. The nests of ant species which carry their seeds occur in the top three zones of these soils: the uppermost litter-humus zone; the grey leached zone immediately beneath; and the brown, nutrient-rich zone which is lowermost (Stickel, 1928; Vessel, 1939; Headley, 1949; Gorman, Newman, Beverage & Hatfield, 1972). Studies by Lyford (1963) have shown that the soil within and around nests is differentiated from its surroundings by ant activity. The ants carry material upwards from the brown zone, which is rich in nitrogen, iron, aluminium, phosphorus, calcium, magnesium and potassium (McCool, Veatch & Spurway, 1923; Niklas, 1924), and moist materials which show strong nitrification properties (Romell & Heiberg, 1931) are carried downwards from the litter-humus zone. In addition, the ants rarely move particles greater than 4 mm in diameter and this fine material has a greater water holding capacity than the surrounding soil (Stickel, 1928). Further studies have demonstrated the accumulation of nutrients and the alteration of the physical structure of soils in and around ant nests (Gentry & Stiritz, 1972; Beattie & Culver, 1977). Ant nests may be safe sites for violet and other myrmecochore seeds, primarily as a result of the particular mix of organic and inorganic nutrients, soil texture and water availability. Headley (1952) and Lyford (1963) estimated that the density of ant nests ranged between 60,000 and 100,000 per hectare in various types of forest. By no means all of these nests belong to ant species which carry seeds. However, their abundance and ubiquity suggests that by attracting ants to its seeds, a plant species has a reliable means of finding safe sites. In habitats where these are limited or patchy this behaviour would be selectively advantageous.

Factors which determine germination and establishment are among the most important influencing the patterning of individuals and populations of plants (Harper & White, 1974). Systems involving seed predators and seed dispersal agents competing for seed crops are well known and often result in the surviving seeds ending up in widely spaced caches (Janzen, 1971). For example, when germination and establishment are successful, the result is the non-random, patchy occurrence of tree seedlings and adults (Abbott & Quink, 1970; Wiens, 1976). The *Viola* data indicate similar mechanisms of plant patterning. The way in which plant individuals and populations are spaced has a profound effect upon gene flow.

GENE FLOW

Gene flow has at least two components in plants: pollen flow and seed flow. The data on pollinator behaviour (Tables 6 and 7) indicate the importance of plant patterning (e.g. spacing) to the pollen component of gene flow. While the correlations between pollinator flight parameters and plant spacing parameters are positive, the remaining data show that as spacing increases or becomes more variable, the number of flights between plants and the quantities of pollen moved are both reduced. This means that with the overall patterning of adult plants being non-random, pollinator service creates a positive feedback, i.e. once clumping is initiated by ant activity, most pollen exchange occurs within clumps. Pollen exchange also increases with greater proximity and regularity of spacing of adults. This is likely to increase the amount of selfing and sibling pollinations (Bateman, 1956). The pollen component of gene flow is clearly restricted in space. Wright (1943), Jain & Bradshaw (1966), Bradshaw (1972) and Endler (1973) have demonstrated in various ways that restricted gene flow promotes inbreeding and the divergence of subpopulations so that differentiation may occur on a very localised scale. Jain (1976) and Selander & Hudson (1976) have emphasised the importance of the process of population subdivision in optimising local fitness while generating between-population variability. This point will be discussed again below.

There are, however, two important processes which produce counter effects. Firstly, pollen dispersal is leptokurtic; small amounts are transported much longer distances, for example, during exploratory flights of larger bees and lepidopterans. Secondly, the mean dispersal distance of seeds is significantly greater than the mean dispersal distance of pollen for all *Viola* species studied (Table 8). Once again most seed dispersal distributions show leptkurtosis. Thus, some pollen and seed dispersal is over longer distances with a greater probability of gene exchange between clumps of plants. The data suggest that the seed component is the major vehicle for gene flow in *Viola*.

It should be noted that the frequency distributions of both pollen and seed dispersal distances generate very large variance values (Table 8). Neighbourhood sizes calculated by the usual basic formula $4\pi\sigma^2 d$ (Wright, 1946) or its derivatives, e.g. $4\pi d (\frac{1}{2}s^2_p + s^2_s)$ (Levin & Kerster, 1968) run into the order of tens of thousands of plants in *Viola*. If seed dispersal distances are sufficient to maintain some gene flow between population subunits, then this result may be realistic. The picture of violet organisation which emerges is large populations subdivided into semi-isolated, mostly inbreeding subunits. However, the next order of priority will be the recalculation of neighbourhood sizes based on formulae which assume non-normal distributions (Wright, 1969).

CHASMOGAMY VERSUS CLEISTOGAMY

At this point it is important to recall that, in addition to seed-set by insect-pollinated (chasmogamous) flowers, many species of *Viola* produce large quantities of seed by means of obligately self-fertilised (cleistogamous) flowers. Grant (1971) and Williams (1975) have pointed out that uniparental sexual reproduction such as cleistogamy gives rise to offspring genetically very similar to the parents, and may differ little from asexual reproduction in this respect. The advantage of cleistogamy is that genes responsible for a local adaptive peak will be brought to a high localised frequency (Stebbins, 1970; Williams, 1975). If this is associated with positive linkage disequilibrium, then cleistogamy will speed response to local selective pressures and generate local spatial fitness and at least short-term temporal fitness (Thompson, 1976). Since cleistogamy, which involves obligate self-pollination, is common in *Viola*, and since the data show buffered gene flow among chasmogamous flowers (see section on pollen vectors), it can be argued that cleistogamy is the optimal mode of reproduction so long as inbreeding depression does not arise and the environment favours existing gene combinations.

When this condition breaks down, however, chasmogamous reproduction may be at a premium. Populations of small to moderate sizes cannot harbour all possible genotypes simultaneously (Thompson, 1976). Therefore, chasmogamy may fulfil the basic sexual function, envisaged by Maynard-Smith (1968), of long(er) distance gene flow between breeding units with different adaptive norms. The kurtosis of both pollen and seed dispersal frequency distributions indicate potential flow between breeding units. Pollination of chasmogamous flowers will generate a small proportion of novel recombinants. The ballistic data show that some will remain close to their parents while others will travel to other breeding units. If selection favours new combinations of alleles, or new alleles, chasmogamy is a mechanism for tapping the resources of between-colony or between-population variation. Allard, Jain & Workman (1968) and Jain (1969) have shown that predominantly self-pollinating species such as *Avena barbata* harbour a remarkable amount of genetic variability which, although distributed among large numbers of monomorphic breeding units, may be released by means of low frequency outcrossing. This is a population structure which pollen and seed vectors appear to impose upon some species of *Viola*. Gene flow frequency distributions are leptokurtic and positively skewed indicating high inbreeding coefficients with low frequency outcrossing. It should be noted that Schaal (1976) has found almost as much within-breeding unit (neighbourhood) genetic diversity as within-population genetic diversity in *Liatris cylindracea*. However, since neighbourhoods also appear to be markedly differentiated the function of chasmogamy suggested here remains much the same.

This interpretation of the role of chasmogamy and gene flow indicates that its function is to speed evolutionary responses to changing selection pressures by promoting both gene migration and reassortment. The intriguing possibility remains, however, that for some *Viola* species in particular, chasmogamy functions to slow or moderate evolutionary response (Thompson, 1976). This possibility may be envisaged for a violet population reproducing primarily by cleistogamous seed (Clausen *et al.*, 1964). A population such as this may generate substantial linkage disequilibrium (Allard, Babbel, Clegg & Kahler, 1972; Clarke, 1974) which under some environmental regimes reduces fitness (Levins, 1968). Adaptation may become excessively 'finely tuned' resulting in an inflexible genome incapable of response to changing selection pressures. Chasmogamy either within or between breeding units would counter this effect by disrupting linkage disequilibria.

ACKNOWLEDGEMENT

Much of this research was supported by National Science Foundation Grant DEB 76–02227 to the author and to David Culver; this is gratefully acknowledged.

REFERENCES

ABBOTT, H. G. & QUINK, T. F., 1970. Ecology of eastern white pine seed caches made by small forest mammals. *Ecology, 51:* 271–278.

ALLARD, R. W., JAIN, S. K. & WORKMAN, P. L., 1968. The genetics of inbreeding populations. *Advances in Genetics, 14:* 55–131.

ALLARD, R. W., BABBEL, G. R., CLEGG, M. T. & KAHLER, A. L., 1972. Evidence for co-adaptation in *Avena barbata. Proceedings of the National Academy of Sciences of the U.S.A., 69:* 3043–3048.

BATEMAN, A. J., 1956. Cryptic self-incompatability in the wallflower, *Cheiranthus cheiri* L. *Heredity, 10:* 257–261.

BEATTIE, A. J., 1969. Studies in the pollination ecology of *Viola*. 1. The pollen content of stigmatic cavities. *Watsonia, 7:* 142–156.

BEATTIE, A. J., 1971. Pollination mechanisms in *Viola. New Phytologist, 70:* 343–360.

BEATTIE, A. J., 1972. Insect visitors to three species of violet (*Viola*) in England. *Entomologist's Monthly Magazine, 108:* 7–11.

BEATTIE, A. J., 1974. Floral evolution in *Viola. Annals of Missouri Botanical Garden, 61:* 781–793.

BEATTIE, A. J., 1976. Plant dispersion, pollination and gene flow in *Viola. Oecologia, 25:* 291–300.

BEATTIE, A. J. & CULVER, D. C., 1977. Effects of the mound nests of the ant *Formica obscuripes* on the surrounding vegetation. *American Midland Naturalist, 97:* 390–399.

BEATTIE, A. J. & LYONS, N., 1975. Seed dispersal in *Viola* (Violaceae): Adaptations and strategies. *American Journal of Botany, 62:* 714–722.

BRADSHAW, A., 1972. Some of the evolutionary consequences of being a plant. *Evolutionary Biology, 5:* 25–48.

BRATTON, S. P., 1976. Resource division in an understory herb community: responses to temporal and microtopographic gradients. *American Naturalist, 110:* 679–693.

CLARKE, B., 1974. Causes of genetic variation. *Science, 186:* 524–525.

CLAUSEN, J., CHANNEL, R. B. & UZI NUR, 1964. *Viola rafinesquii*, the only Melanium violet native to North America. *Rhodora, 66:* 32–46.

CULVER, D. C. & BEATTIE, A. J., 1978. Myrmecochory in *Viola*: dynamics of seed-ant interactions in some west Virginia species. *Journal of Ecology*

DROZDZ, A., 1966. Food habits and food supply of rodents in the beech forest. *Acta Theriologica, 11:* 363–384.

DYMES, T. A., 1916. The seed mass and dispersal of *Helleborus foetidus. Journal of the Linnean Society* (*Botany*), *43:* 433–455.

EHRLICH, P. R. & EHRLICH, A. H., 1961. *How to Know the Butterflies.* Dubuque: Brown.

EMLEN, J. M., 1973. *Ecology: An Evolutionary Approach.* Menlo Park: Addison Wesley.

ENDLER, J. A., 1973. Gene flow and population differentiation. *Science, 179:* 243–249.

FRANKLAND, J. C., OVINGTON, J. D. & MACRAE, C., 1963. Spatial and seasonal variation in soil, litter and ground vegetation in some Lake District woodlands. *Journal of Ecology, 51:* 97–112.

FRENCLOWA, I., 1962. The isolation of an alkaloid from *Viola odorata* roots. *Acta Poloniae pharmaceutica, 18:* 187–195.

GENTRY, J. B. & STIRITZ, K. L., 1972. The role of the Florida Harvester Ant, *Pogonomyrmex badius*, in old field mineral nutrient relationships. *Environmental Entomology, 1:* 39–41.

GORECKI, A. & GEBCZYNSKA, Z., 1962. Food conditions for small rodents in a deciduous forest. *Acta Theriologica, 6:* 275–295.

GORMAN, J. L., NEWMAN, L. S., BEVERAGE, W. W. & HATFIELD, W. F., 1972. *Soil Survey of Greenbriar County, West Virginia.* U.S.D.A. Soil Conservation Service, West Virginia.

GRANT, V., 1971. *Plant Speciation.* New York: Columbia Univ. Press.

HANDEL, S. N., 1976. Dispersal ecology of *Carex pedunculata* (Cyperaceae), a new north American myrmecochore. *American Journal of Botany, 63:* 1071–1079.

HARPER, J. L., CLATWORTHY, J. N., McNAUGHTON, I. H. & SAGAR, G. R., 1961. The evolution and ecology of closely related species living in the same area. *Evolution, 15:* 209–227.

HARPER, J. L., WILLIAMS, J. T. & SAGAR, G. R., 1965. The behaviour of seeds in soil. I. The heterogeneity of soil surfaces and its role in determining the establishment of plants from seed. *Journal of Ecology, 53:* 273–286.

HARPER, J. L. & WHITE, J., 1974. The demography of plants. *Annual Review of Ecology and Systematics, 5:* 419–463.

HEADLEY, A. E., 1949. A population study of the ant *Aphaenogaster fulva* ssp. *aquia* Buckley. *Annals of the Entomological Society of America, 42:* 265–272.

HEADLEY, A. E., 1952. Colonies of ants in locuts woods. *Annals of the Entomological Society of America, 45:* 435–442.

HIGGINS, L. G. & RILEY, N. D., 1970. *Butterflies of Britain and Europe.* London: Collins.

HORHAMMER, L., WAGNER, A., RASPRIM, R., MABRY, T. J. & ROSLER, H., 1965. Uber die Struktur never und bekannter Aavon-C-Glykoside. *Tetrahedron Letters*, No. 22: 1707–1711.

JAIN, S. K., 1969. Comparative ecogenetics of two *Avena* species occurring in central California. *Evolutionary Biology, 3:* 73–118.

JAIN, S. K., 1976. Patterns of survival and microevolution in plant populations. In S. Karlin & E. Nevo (Eds), *Population Genetics and Ecology*. New York: Academic Press.

JAIN, S. K. & BRADSHAW, A. D., 1966. Evolutionary divergence among adjacent plant populations. 1. The evidence and its theoretical analysis. *Heredity, 21:* 407–441.

JANZEN, D. H., 1970. Herbivores and the number of tree species in tropical forests. *American Naturalist, 104:* 501–528.

JANZEN, D. H., 1971. Seed predation by animals. *Annual Review of Ecology and Systematics, 2:* 465–492.

LEVIN, D. A. & KERSTER, H. W., 1968. Local gene dispersal in *Phlox*. *Evolution, 22:* 130–139.

LEVIN, D. A. & KERSTER, H. W., 1969. The dependence of bee-mediated pollen and gene dispersal upon plant density. *Evolution, 23:* 560–571.

LEVIN, S. A., 1976. Population dynamic models in heterogeneous environments. *Annual Review of Ecology and Systematics, 7:* 287–310.

LEVINS, R., 1968. *Evolution in Changing Environments.* Princeton: Univ. Press.

LYFORD, W. H., 1963. Importance of ants to brown podzolic soil genesis in New England. *Harvard Forest Paper, 7:* Cambridge: Harvard University Press.

MARTIN, A. C., ZIM, H. S. & NELSON, A. L., 1951. *American Wildlife and Plants: A Guide to Wildlife Food Habits.* New York: Dover.

MAYNARD-SMITH, J., 1968. Evolution in sexual and asexual populations. *American Naturalist, 102:* 469–473.

McCOOL, M. M., VEATCH, J. O. & SPURWAY, C. H., 1923. Soil profile studies in Michigan. *Soil Science, 16:* 95–106.

NIKLAS, H., 1924. Untersuchungen über Bleichsand und Orterdebildungen in Waldboden. *Internationale Mitteilungen für Bodenkunde, 14:* 50–54.

PARIS, R., 1963. The distribution of plant glycosides. In T. Swain (Ed.), *Chemical Plant Taxonomy*. New York: Academic Press.

PIANKA, E. R., 1974. *Evolutionary Ecology*. New York: Harper & Row.

PLATT, W. J., 1976. The natural history of a fugitive prairie plant (*Mirabilis hirsuta* (Pursh) MacM.). *Oecologia, 22:* 399–409.

RABOTNOV, T. A., 1969. On coenopopulations of perennial herbaceous plants in natural coenoses. *Vegetatio, 19:* 87–95.

REISE, K. & WIEDEMANN, G., 1975. Dispersion of predatory forest floor arthropods. *Pedobiologia, 15:* 106–128.

RICKLEFS, R. E., 1973. *Ecology*. Newton, Mass.: Chiron Press.

RIDLEY, H. N., 1930. *The Dispersal of Plants Throughout the World*. Ashford, England: L. Reeve.

ROMELL, L. G. & HEIBERG, S. O., 1931. Types of humus layer in the forests of north-eastern United States. *Ecology, 12:* 567–608.

SCHAAL, B. A., 1976. Genetic diversity in *Liatris cylindracea*. *Systematic Botany, 1:* 163–168.

SELANDER, R. K. & HUDSON, R. O., 1976. Animal population structure under close inbreeding: the land snail *Rumina* in southern France. *American Naturalist, 110:* 695–718.

STEBBINS, G. L., 1970. Variation and Evolution in Plants: Progress during the past twenty years. In M. K. Hecht & W. C. Steere (Eds), *Essays in Honor of Theodosius Dobzhansky*. Englewood Cliffs, New York: Appleton-Century-Crofts.

STICKEL, P. W., 1928. Physical characteristics and silvicultural importance of podzol soil. *Ecology, 9:* 176–190.

STICKEL, L. F. & WARBACH, O., 1960. Small mammal populations of a Maryland woodlot. *Ecology, 41:* 269–286.

STRAKSBAUGH, P. D. & CORE, E. L., 1970. Flora of West Virginia, *1. West Virginia University Bulletin*, Ser. 70.

STRUIK, G. L. & CURTIS, J. T., 1962. Herb distribution in an *Acer saccharum* forest. *American Midland Naturalist, 68:* 285–296.

THOMPSON, V., 1976. Does sex accelerate evolution? *Evolutionary Theory, 1:* 131–156.

VESSEL, A. J., 1939. Soils of Greenbrier County. In P. H. Price & E. T. Heck (Eds), *West Virginia Geological Survey, Greenbriar County*. Wheeling, West Virginia.

WEINS, J. A., 1976. Population responses to patchy environments. *Annual Review of Ecology and Systematics, 7:* 81–120.

WILLIAMS, G. C., 1975. *Sex and Evolution.* Princeton: Univ. Press.

WITHERBY, H. F., JOURDAIN, F. C. R., TICEHURST, N. F. & TUCKER, B. W., 1952. *The Handbook of British Birds.* London: Witherby.

WRIGHT, S., 1943. Isolation by distance. *Genetics, 28:* 114–138.

WRIGHT, S., 1946. Isolation by distance under diverse systems of mating. *Genetics, 31:* 39–59.

WRIGHT, S., 1969. *The Evolution and Genetics of Populations.* 2. *The Theory of Gene Frequencies.* Chicago: Univ. of Chicago Press.

YOUNG, V., 1934. Plant distribution as influenced by soil. *Ecology, 15:* 154–196.

PLATES 7 TO 23

PLATE 7

An insect visit to an anemophilous plant: a syrphid fly (Diptera: Syrphidae) *Melanostoma* sp. feeding on pollen of the sedge *Carex binervis*, Devon, 1970. In this instance monoecism makes pollination by such visits improbable. Compare *Plantago*, Stelleman p. 42.

PLATE 8

Honeybee (*Apis mellifera*) collecting nectar from a ♀ catkin of the common sallow (*Salix cinerea*), Devon, 1969 (see A. D. J. Meeuse, p. 48).

PLATE 9

Division of labour in the florets of a composite. *Bombus pratorum* feeding from the bisexual central florets of the greater knapweed (*Centaurea scabiosa*). Note pollen borne on the forked stigma. The outer ring of more showy florets are neuter, although here they can be seen to be dusted with pollen, presumably as a result of insect activity (see van der Pijl, p. 83).

PLATE 10

Florets of *Solidago virgaurea* showing apical fusion of the stigma branches, leaving a narrow basal "lantern slit", the margins of which become receptive late in anthesis. As in many Compositae, the fused pistil acts as a "secondary stamen", presenting the pollen (see van der Pijl, p. 82).

PLATE 11

A "*Solanum*"-type flower (see Vogel, p. 91). *Bombus lucorum* hangs from the conspicuous yellow anther cone of *Solanum dulcamara*. This is a species in which anthers share feeding and donor functions, and all are yellow and of the same size.

PLATE 12

Convolvulus arvensis, an oligandrous sympetalous actinomorph of a simple type (see Vogel, p. 90) being visited by a solitary bee (*Halictus* sp.). Alderney, Channel Islands, July 1969.

PLATE 13

A beetle- and ichneumon-pollinated orchid. A. Skipjack beetle *Athous haemonboidalis* pollinating Twayblade, *Listera ovata*, Devon, 1968 (see Proctor, p. 108). Note pollinia on head of beetle are of a primitive form. B. An ichneumon fly visits the same species, Dorset, 1968.

PLATE 14

Orchid pollination. A yellow dung-fly (*Scopeuma stercorianum*) carrying pollinia of the heath spotted orchid (*Dactylorhiza maculata* subsp. *ericetorum*) upon its head. The pollinia have come to a "forward" position, ready for insertion. Devon, 1968.

PLATE 15A

Flower of *Silene nutans* (Caryophyllaceae), possessing a typical syndrome suitable for pollination by night-flying moths (see Proctor, p. 108) with whitish flowers which open and are scented for only three successive nights, five stamens ripening on each of the two first nights and the styles ripening on the third. The photograph is taken on the second night. Viscous stalked glands, which probably act as extrafloral nectaries to distract "minor visitors" are clearly visible. Devon, June 1969.

PLATE 15B

A Silver-Y moth (*Plusia gamma*) visits the night-scented pale flowers of honeysuckle (*Lonicera periclymenium*) (see Proctor, p. 108). A similar photograph formed the basis for the motif of the conference. Devon, 1971.

PLATE 16

Three examples of specialised zygomorphic flowers in the Labiatae; their differences probably reflect a combination of selection pressures for reproductive isolation and for maximum efficiency of pollination by incompletely constant pollen vectors (see Proctor, p. 108). A. *Lamium album* with vertical access and a broad tube, showing bilocular anthers and style included in the hood. B. *Salvia glutinosa*, visited by *Bombus agrorum*, showing horizontal access with well-marked landing platform and narrow tube, with exserted style and unilocular anthers. Note small unspecialised visitor (extreme left) to viscous stalked glands on the sepals (extra-floral nectaries). C. *Lamiastrum galeobdolon*, visited by *Bombus*? *agrorum*, with downwards access into a constricted "S"-bend tube with a ring of hairs, and a reduced landing-platform; anthers bilocular, included with style.

PLATE 17

In contrast to Plate 16, a "zygomorphic" platform and tube type syndrome suitable for pollination by bees, similar to that in the Labiatae has evolved in *Iris* (Iridaceae) in which each "flower" unit is an outer perianth segment ("fall"), overlaid by a petaloid stigma to form a tube. *Iris pseudacorus*, visited by *Bombus hortorum*. Note the full corbicula on the bee's leg.

Continued overleaf

(*Facing p. 164*)

PLATE 18

Sternotribic zygomorphy (see Proctor, p. 108) illustrated by the half-flower of *Aconitum anglicum* (Ranunculaceae)—a typical specialised bee-flower. Note the large inflated posterior sepal with in the roof, a long-clawed tubular nectary. This protandrous flower is in the earlier (male) phase. Devon, June 1969.

PLATE 19

Unspecialised pollinators. A. Dolichopodid flies (Diptera, Dolichopodideae) visiting a flower of *Potentilla reptans* (Rosaceae), which shares in a Muellerian manner a simple polyphilic syndrome with other common northern temperate grassland genera (*Ranunculus, Helianthemum,* etc.) (see Proctor, p. 109). Alderney, Channel Islands, 1969. B. The hoverfly *Sphaerophoria? rueppellii* ♂ on tormentil, *Potentilla erecta*, Devon, 1968. C. A sawfly visits *Ranunculus acris*. Note the abundant pollen on the front of the insect.

PLATE 20

Two examples of visits by hoverflies to unspecialised flowers in clusters or umbels (see Proctor, p. 109). A. *Helophilus pendulus* on hawthorn (*Crataegus monogyna*), Devon, 1969. B. *Scaeva pyreasti* on hogweed (*Heracleum sphondylium*). Devon, 1968.

PLATE 21

Pollination of an alien by a rarity. Red valerian (*Kentranthus ruber*) from southern Europe, a specialised butterfly-pollinated plant, visited by a male Lulworth Skipper butterfly (*Thymelicus acteon*), a primarily south European species with a few relict sites on the south coast of England. Dorset, 1975 (see Valentine, p. 121). Note pollen on the head of the butterfly.

PLATE 22

Legitimate and illegitimate access. A. *Bombus? agrorum* visits the pendulous bee-flowers of the bluebell *Endymion non-scriptus*. B. The honey bee, *Apis mellifera*, "robs" a flower of *Endymion hispanicus* by probing between the perianth segments at the base of the flower to obtain easy access to nectar.

PLATE 23

Bumblebee visiting *Impatiens glandulifera*, Dorset, 1972 (see Valentine, p. 119).

(All photographs are by M. C. F. Proctor.)

PLATE 7

PLATE 8

PLATE 9

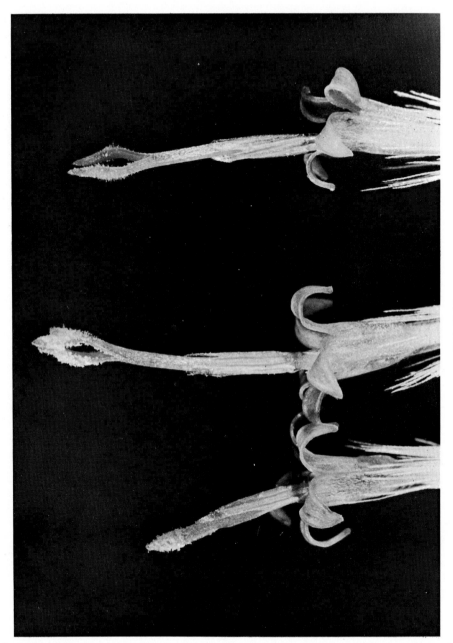

PLATE 10

PLATE 11

PLATE 12

PLATE 13A

PLATE 13B

PLATE 14

PLATE 15B

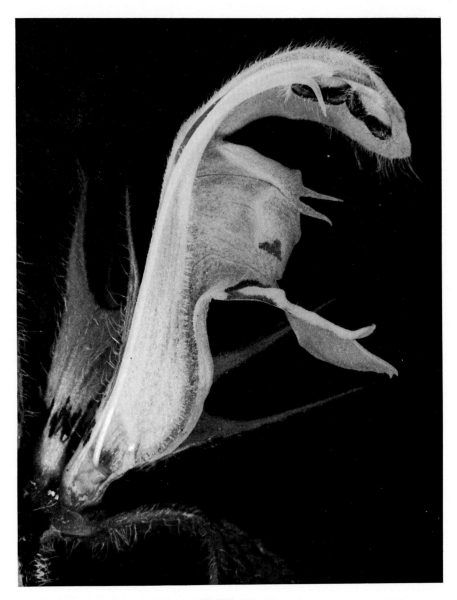

PLATE 16A

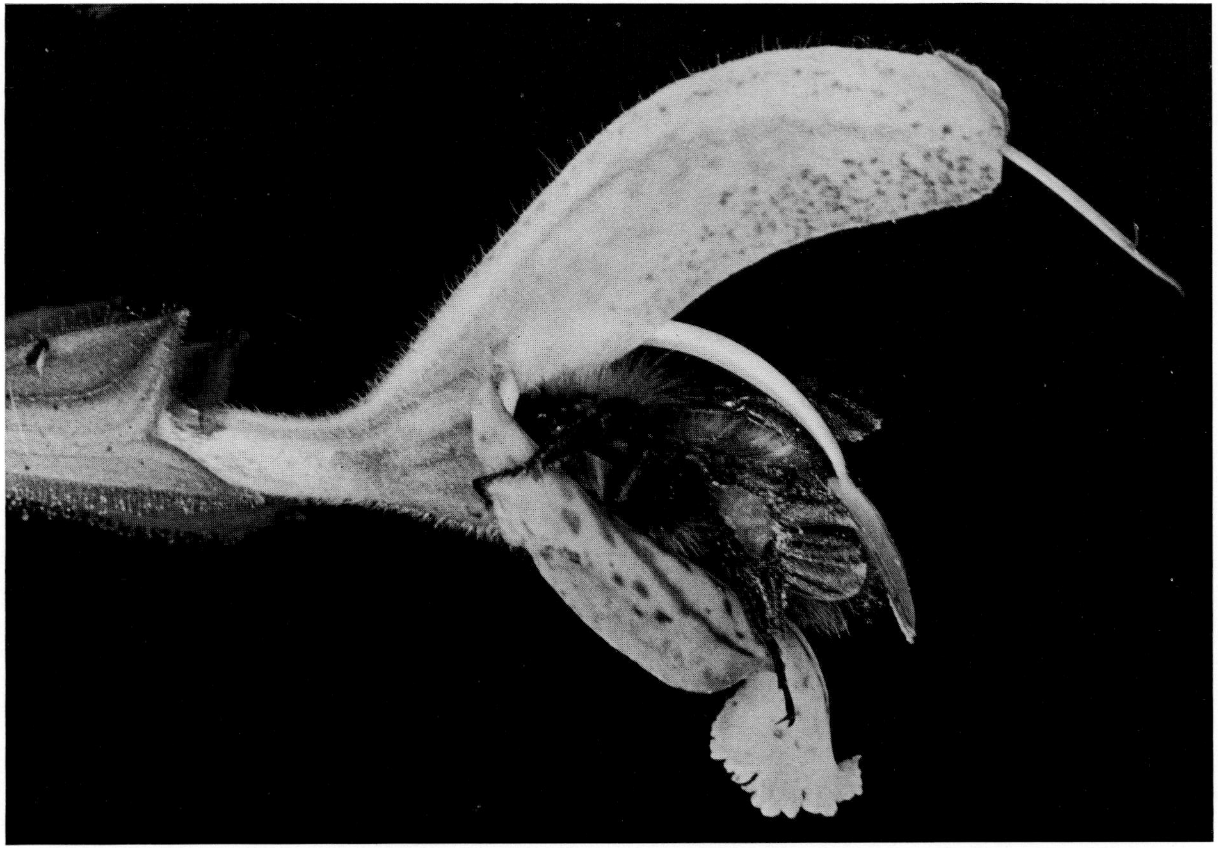

PLATE 16B

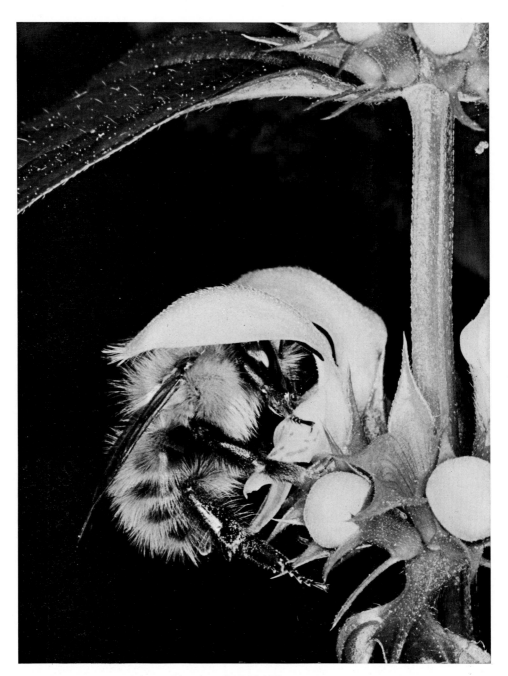

PLATE 16C

PLATE 17

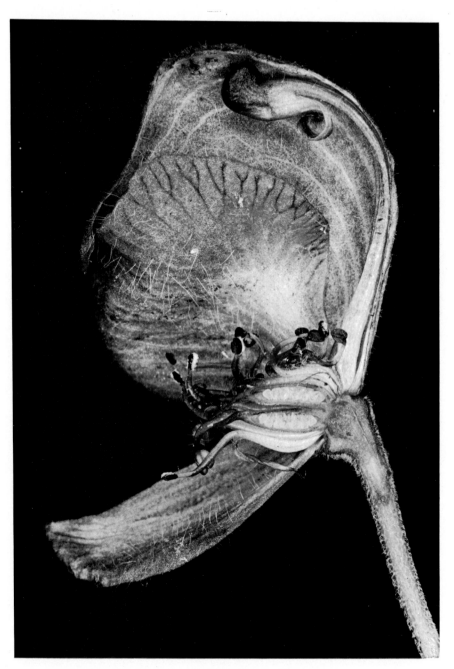

PLATE 18

PLATE 19A

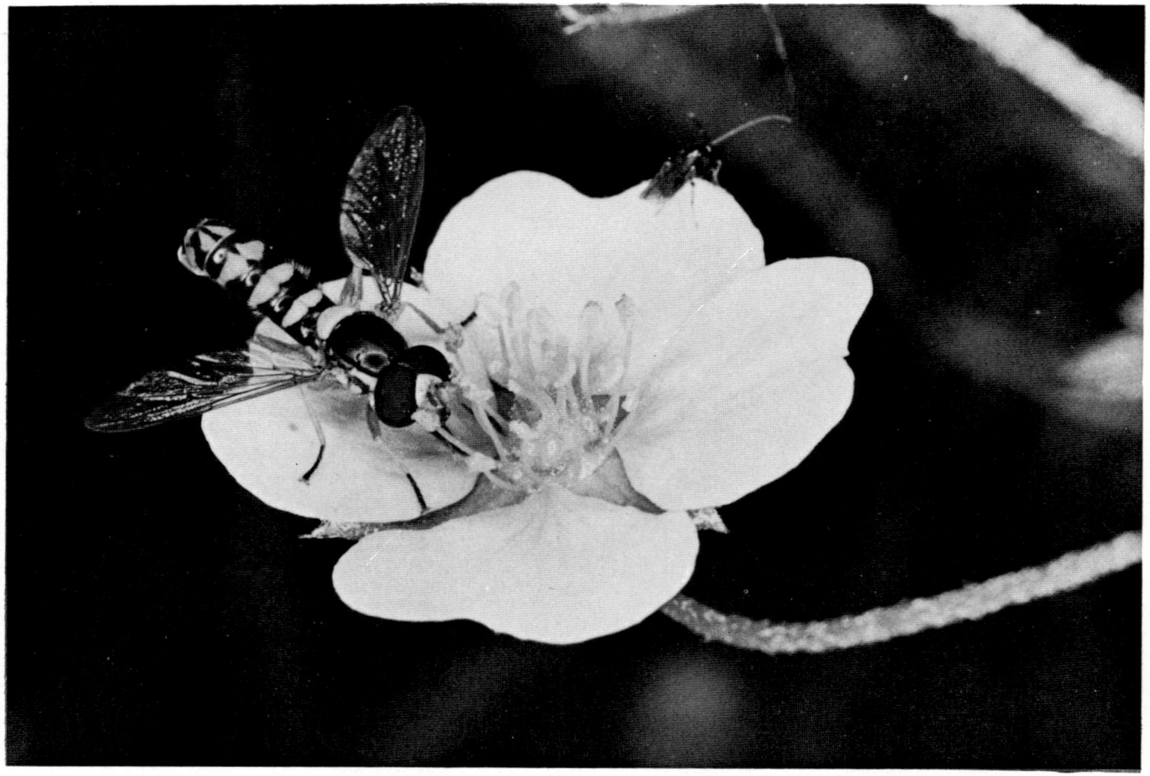

PLATE 19B

PLATE 19C

PLATE 20A

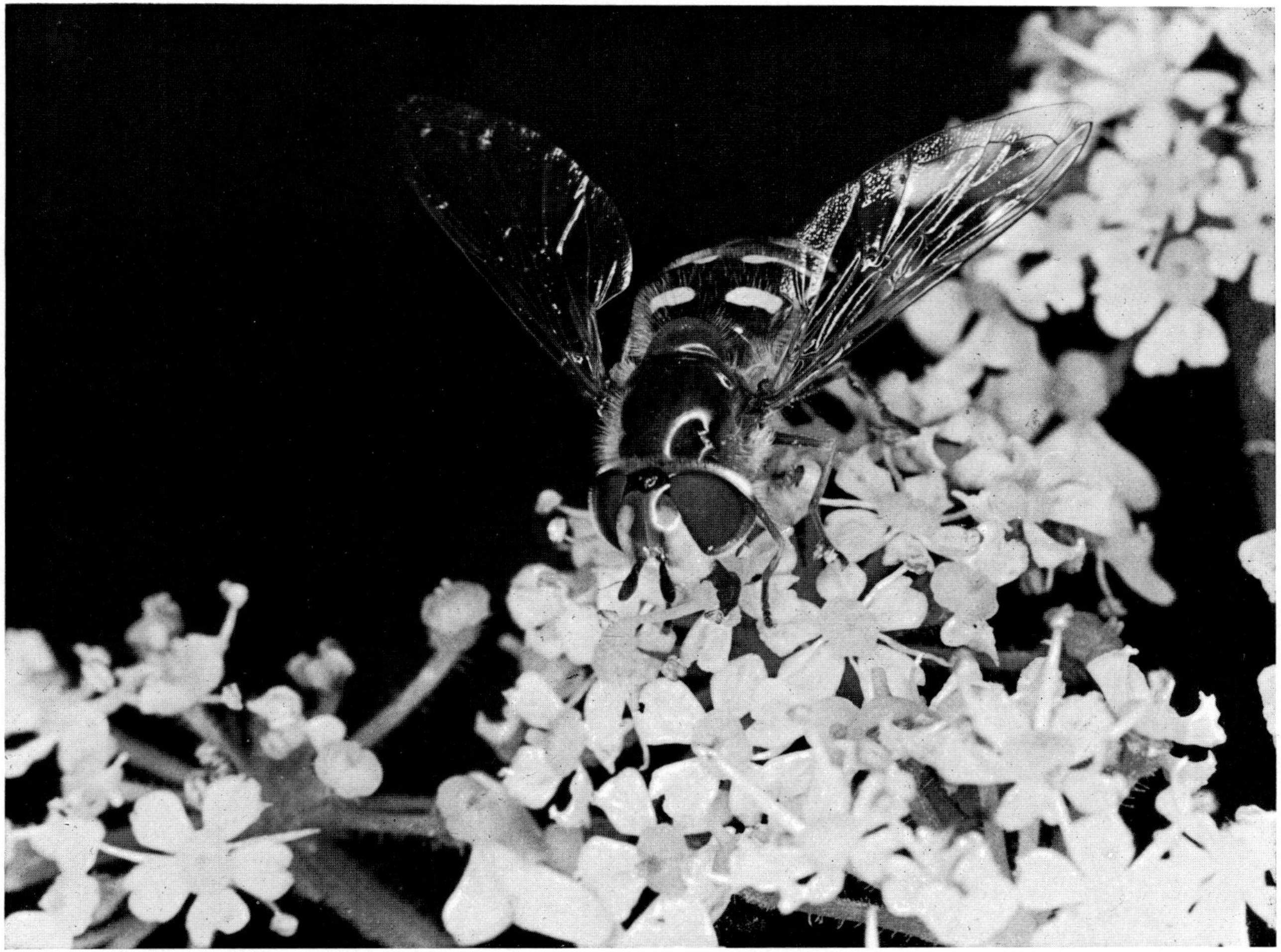

PLATE 20B

PLATE 21

PLATE 22A

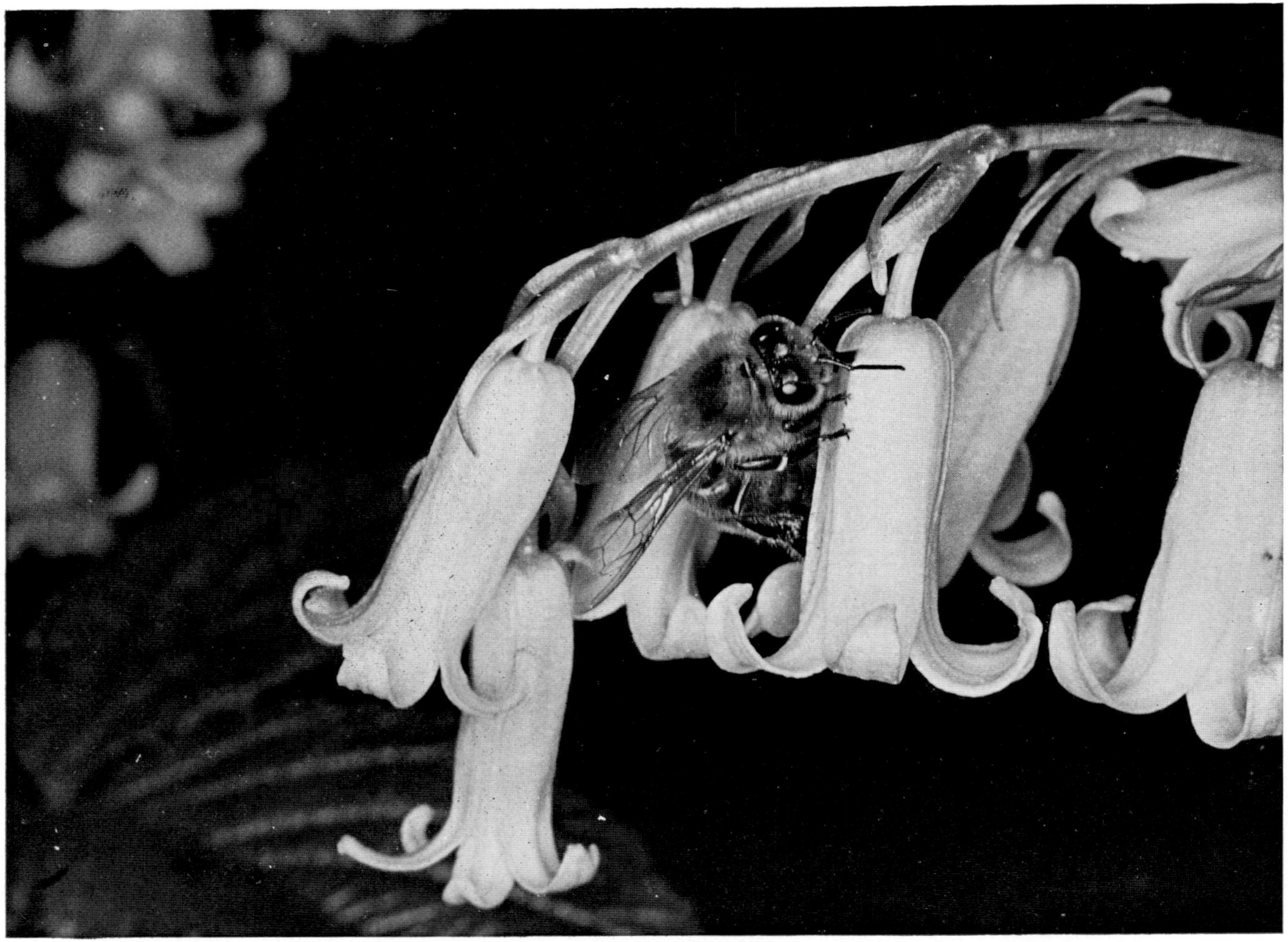

PLATE 22B

PLATE 23

Estimation of neighbourhood size in two populations of *Primula veris*

A. J. RICHARDS AND HALIJAH IBRAHIM

Department of Plant Biology, University of Newcastle upon Tyne, U.K.

Attempts were made to estimate neighbourhood size in two large Northumberland populations of *Primula veris*, an outbreeding perennial with a dimorphic incompatibility system. Seed travel was determined using sticky tape; two estimations gave similar means of 0.11 and 0.12 m, with variances of 0.67 and 1.77. Indirect estimates using the position of Fl hybrids gave rather larger means, and smaller variances.

Pollen travel was estimated by counting legitimate pollen grains on stigmas of plants, the distance from which to the nearest legitimate pollinator was known. Mean travel varied from 1.11 to 2.63 m, with 95% of the grains travelling from 3.5 to 12 m. Estimates of variance ranged from 1.7 to 13.1. Distributions were leptokurtic, answering the equation $y=x/a+bx$; this is thought to be due to forage and escape flights of pollinators.

Density of available pollen donors proved to be very variable, ranging from 0.25 to 6.25 clones per m². As estimations of pollen travel are also density dependent, density was considered, directly and indirectly, to contribute the most variability to estimates of neighbourhood size, which ranged from 1 to 562 individuals. Considerations of constraints on the estimates suggest that neighbourhood size values in the populations studied might lie between 5 and 200, with an average of about 30.

It is emphasised that although small neighbourhoods may significantly affect heterozygosity in the seedling generation, overlap in time and space will largely negate inbreeding effects in subsequent generations, when they occur in large populations. It is stressed that plant populations may often be sufficiently small to allow inbreeding effects, such as a reduction of heterosis, and genetic drift, but that further work is required to demonstrate this.

KEY WORDS:—*Primula*—heteromorphy—pin—thrum—pollen-travel—seed-travel—neighbourhood.

CONTENTS

INTRODUCTION

It is widely accepted that population size may be of evolutionary importance, in that inbreeding effects, resulting from the lowering of heterozygosity below that expected in an outbreeding population, will occur in small populations. As a result, reduction of heterosis may restrict vigour in small populations, and poly-morphic genes may become fixed in the homozygous state independently of selection pressures (Wright, 1938, 1943, 1946, 1948).

It seems that plant populations may be very small, indeed much smaller than is immediately apparent, due to clonality (e.g. Harberd, 1961, 1962) and poor dispersal of pollen and seeds (Levin & Kerster, 1968). However, very few estima-tions of population size in perennial plants have been published, and attempts to obtain such figures can be complicated by clonality, and the determination of seed and pollen dispersal.

It is important to distinguish between the number of individuals potentially exchanging genes in one season (which may influence the heterozygosity of seedlings arising from one season's seed), and the number of individuals which may potentially exchange genes irrespective of time (which will be larger, and thus tend to negate reduction of heterozygosity in a parental population). The latter estimation will require data on dormant seeds, non-flowering individuals and generation and population overlap, and may be very difficult to achieve.

The present paper reports attempts to estimate the former, using the concept of neighbourhood size (Kerster, 1964), which has been developed from Wright (1943) and Dobzhansky & Wright (1943), and has been subsequently employed by Lamotte (1951) and Levin & Kerster (1968). The choice of the cowslip, *Primula veris* was governed by a desire to use an outbreeding perennial plant with varying population sizes: the peculiar breeding system with dimorphic pollen has aided estimation of pollen travel.

ESTIMATION OF POLLEN TRAVEL

Although there exists a good deal of evidence concerning the dispersal of wind-distributed pollen, summarised in Proctor & Yeo (1973), measurements of the travel of insect-carried pollen are fewer, and mostly rely on indirect estimates of vector flight-distances (Bateman, 1946, 1947; Free, 1962, 1970; Levin & Kerster, 1968), or of fruit and seed-set on isolated plants (Bateman, 1947; Free, 1962). Only in Levin (1972) has the actual dispersal of insect-carried pollen been estimated, and this concerns relatively rare occurrences of interspecific pollinations between *Phlox divaricata* and *P. bifida*.

Another technique has been to use dominant gene markers in seedlings (Stephens & Finker, 1953), and this has the advantage of combining pollen and seed travel. In all known cases in cotton, however, marker genes are pleiotropically associated with factors of growth and vigour, and this may be a general problem.

In order to obtain direct estimates of pollen dispersal, it is necessary to have a natural or artificial marker. Experiments using methylene blue powder (Stephens & Finker, 1953) or fluorescent powder (Woodell, pers. comm.) can be successful, but lead to problems of detection, dilution by rain or dew, pollinator bias, and poor adherence of the powder. Pilot experiments using pollen marked with a radio-active tracer have failed, due to the very small quantities of pollen involved in cross-pollination, and difficulty in obtaining successful uptake of tracer by pollen in the anther.

It is likely that natural pollen markers will be more useful, as used by Levin (1972) in *Phlox*. However, intraspecific variability in pollen is rare and seems to be limited to dimorphic plants (usually associated with diallelic sporophytically controlled incompatibility systems), as in many *Primulaceae*, *Armeria*, etc. (Lewis, 1954).

In most dimorphic systems, two cross-compatible, but largely self-incompatible genotypes (Ss and ss) exist in roughly equal numbers. It is usual to find that these are associated with physical characteristics of the flower which encourage legitimate and discourage illegitimate pollinations, and this association is maintained by a linkage group (Dowrick, 1956).

In *Primula*, and several other genera in the Primulaceae (*Hottonia, Douglasia, Androsace, Dionysia*), as well as in some other genera (*Oxalis, Linum, Forsythia*), it is well known that a heterostyly occurs, in which one genotype (Ss) has anthers adherent to the mouth of the corolla tube and a short style, with the stigma inserted well down the tube (thrum). The other (ss) has a long style with a stigma which is usually slightly exerted, but in which the anthers occur well inside the corolla tube (pin). In all the heterostylous Primulaceae, the pin stigma is round with long papillae, and the pollen is small (*c.* 25 μm in diameter in *P. veris*), whereas the thrum stigma is flattened with short papillae (or almost smooth) and the pollen is much larger (*c.* 45 μm). There seems little doubt that these features encourage legitimate pollination, although illegitimate pollination is a striking feature, especially of pin flowers.

The value of this system to the present discussion is that a balanced pollen marker is found in every population, and that the proportion of legitimate and illegitimate pollinations is readily determined by microscopic examination of the stigma.

Method

Individual flowers which were fully open, but in which the stigma was still green and turgid were collected in individual polythene bags and sealed. In each case, the distance to the nearest clone of the other morph (pin or thrum) was noted. Samples were frozen until they could be examined. Stigmas were excised and mounted on a slide under a coverslip in a drop of cotton blue, and gently squashed. When examined under a microscope, 'foreign' (legitimate) pollen grains could be readily distinguished by size. There is no overlap in size of the pollen of the two morphs. The total number of foreign pollen grains are counted per stigma. This technique is open to two major biases:

(1) The length of time that a flower will have been available for cross-pollination will vary at the time of collection; consequently, values obtained may be low. However, in mature flowers legitimate pollinations may be observed which arrived on the stigma after it ceased to be receptive to fertilisation.

(2) The distance to source of legitimate pollination is calculated by the distance to the nearest potential source. However, a grain may have travelled from a further plant of the same morph. Thus, all readings are potentially underestimates. In fact, the majority of readings were made from plants which were rather isolated, and for which alternative sources were much further distant. Only for short distances, calculated from plants growing in relatively dense stands, may the underestimate be serious.

However, isolated plants may receive rather fewer insect visits than those in denser stands, resulting in a further bias (Levin & Kerster, 1969), although pollen travel to such isolated plants is likely to be further. Thus the data is likely to overemphasise the leptokurtosis of the pollen dispersal.

The assumption is made that the foreign pollen source will be one plant only. In many cases this is manifestly unlikely, and the bias is likely to be unequal if one type is more frequent in the locality, which is usually true of pins as here (Table 1) (Bodmer, 1960), presumably due to the capacity of pins to show some self-fertilisation (Crosby, 1949); but occasionally of thrums (Lees, 1971). It is of interest that the total self-incompatibility exhibited by thrums in *P. vulgaris* (Crosby, 1949) and *P.* × *polyanthus* (Richards, unpubl.) is not true of *P. veris*, although very little thrum × thrum seed is set.

Table 1. Ratios of pins and thrums

Taxon	Population	Clones			Rosettes		
		P	T	%P	P	T	%P
P. veris	Whittle Dene, Dam 1	133	117	53.2	69	61	53.1
P. vulgaris	Whittle Dene, Dam 1	219	205	51.7	342	343	50.0
P. veris × *P. vulgaris*	Whittle Dene, Dam 1 +Dam 2	96	86	52.6	210	218	49.1

Two populations have been studied in Northumberland, chosen for ease of access, lack of disturbance, and contrasting size. At Whittle Dene (45/065685) a large population, of the order of 20,000 individuals, occurs on reservoir embankments, built in 1838. The population occupies a narrow zone around the greater part of two reservoir basins, totalling some 1200 m × 20 m. *P. vulgaris* is also present in more shaded and wetter sites, and many hybrids occur in areas where the parents meet.

At Park Shield (35/893689) a much smaller population, of about 500 individuals occurs in an old limestone quarry: *P. vulgaris* is absent here.

Results

The population around one of the basins at Whittle Dene (Dam 1) has been studied on three consecutive years (1974–6). Analysis of the pollen dispersal by general multiple regression (programme FAFAA, UDF9 computer, University of Oxford) shows a markedly leptokurtic distribution (Fig 1), best fitted by the equation $y = x/a + bx$. The number of legitimate pollen (y) on a stigma is apparently dependent on distance (x), modified by a variable (a) independent of distance and another (b) dependent on distance. It is likely that (b) represents pollinator forage flights and (a) pollinator escape flights (Bateman, 1947).

A summary of the results (Table 2) shows that the mean number of legitimate pollen per stigma varies from 9.8–17.5. Differences between pins and thrums are apparently related to the greater frequency of pins (Table 1). The mean distance of travel varies from 1.11 to 2.63 m, and does not vary much, or consistently between pins and thrums.

In 1976, when sampling occurred after a spell of fine, warm weather, pollen travel was notably further. There is a good deal of variation in the distance travelled by 95% of the grains (from 3.5 to 12 m). This figure is very dependent on the number of isolated samples taken. The mean variance of pollen travel is rather

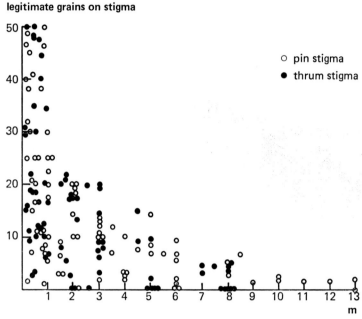

Figure 1. Legitimate (pin on thrum or thrum on pin) pollen grains on stigmae of *P. veris* at Whittle Dene, Dam 1 in 1975. Distances are to the nearest individual of the other morph. Both the leptokurtic distribution and the density dependence of visit frequency are demonstrated.

Table 2. Pollen travel in *P. veris*

Population/ year	No. stigmas P	T	Total legitimate pollen on stigmas P	T	Mean legitimate pollen on stigma P	T	Mean travel onto stigma (m) P	T	Travel of 95% grains onto stigma (m) P	T	Variance pollen travel onto stigma (m)
Whittle Dene Dam 1, 1974	45	30	490	702	16.3	15.6	1.58	1.41	3.5	5	1.82
Whittle Dene Dam 1, 1975	39	35	500	479	12.8	13.6	1.70	1.48	5	4	3.37
Whittle Dene Dam 1, 1976	48	42	492	638	10.2	17.5	2.63	2.61	10	12	13.07
Park Shield 1976	30	24	295	388	9.8	16.1	1.11	1.72	3.7	5	1.75
Whittle Dene Dam 1, 1976 hybrid technique							2.80				5.45
Whittle Dene Dam 2, 1976 hybrid technique							0.98				0.73

consistent in three samples (1.75 to 3.37 m), but in Whittle Dene 1976, the variance is much higher (13.07). This seems to reflect a genuine increase in pollen travel, and hence probably of pollinator activity, rather than more assiduous sampling of isolated plants, which were standardised between experiments as far as possible.

ESTIMATION OF SEED TRAVEL

Method

Isolated plants bearing fruiting spikes with mature seed ready for dispersal were surrounded in the field with 4–6 clear sticky tapes ('sellotape') radiating 5 m from the plant. These were pinned to the ground with thin metal hoops. Although some disturbance resulted from cattle at Park Shield, most tapes stayed intact and in position, and were removed and seeds counted and measured to source after three weeks. In all, 174 seeds at Park Shield and 24 at Whittle Dene were recorded (Table 3). Although the mean travel at the two sites was very similar (0.12 and 0.11 m), the variance was much greater at Park Shield, doubtless due to the larger and more realistic sample size.

This means of sampling seed dispersal is probably quite accurate, as seeds are unlikely to blow or roll once grounded in dense grass sward. It is unlikely that animal carry is important, as mature seeds of *P. veris* lack the gelatinous projection of the seed-coat attractive to ants that is found in *P. vulgaris*.

Table 3. Estimation of seed travel

Population/technique	n	Mean (m)	Variance (m)
Park Shield tapes	170	0.12	1.77
Whittle Dene Dam 1, tapes	24	0.11	0.67
Whittle Dene Dam 1, hybrids	54	0.54	0.31
Whittle Dene Dam 2, hybrids	38	0.43	0.10

ESTIMATION OF POLLEN AND SEED TRAVEL USING HYBRIDS

Due to the frequent hybridisation of *P. veris* with *P. vulgaris* at Whittle Dene, it was possible to use another means of obtaining a rough estimate of pollen and seed travel. Experience has shown that the use of morphological and pollen fertility criteria makes the identification of presumptive Fl hybrids relatively straightforward. Furthermore, analysis of hybrids suggests that F2 and backcross plants may be relatively unimportant, probably due to impaired fertility in the F1 hybrids (Valentine, 1975). The assumptions are made that the parents of the hybrid still survive, that the nearest parent is the presumptive seed parent, and that the nearest plant of the other parent to the presumptive seed parent is the presumptive pollen parent. These assumptions are most likely to be accurate in areas of low density, and these were especially selected. Although this may give an overestimate of pollen and seed travel, this is likely to be countered by the enhanced pollinator travel in low density areas, and the likelihood of more distant parents being involved. The technique is open to serious bias however.

This is clear in estimations of seed travel (Table 3) in which means are higher and variances lower than in the probably reliable estimates from sticky tapes. In estimations of pollen travel (Table 2), the two estimates vary, but scarcely outside the range of figures obtained by the other, also admittedly indirect, technique.

NEIGHBOURHOOD SIZE

This is calculated according

$$NS = 4\pi(\tfrac{1}{2}s^2_p + \tfrac{1}{2}s^2_s)d$$

where s^2_p is the variance of pollen dispersal (m),
s^2_s is the variance of seed dispersal (m),
d is the density (here of flowering clones) per m²,
and follows Kerster (1964), Levin & Kerster (1968) and Levin (1978). A major problem arises from attempts to use variance estimates on leptokurtic distributions. Levin & Kerster (1968) suggest that this distribution results from the superimposition of two normal curves with similar means, but very different variances, these being forage and escape flights of the pollinator. It seems unlikely that the distribution of these flight types do have similar means, but I feel that this suggestion may be otherwise essentially correct. However, they suggest that escape flights, accounting for only 2% of pollinator moves, can be ignored, which seems unfortunate as escape flights may be important when gene dispersal is considered, possibly involving distances between subpopulations with different adaptive modes. Unlike Levin & Kerster, we include all data on pollen dispersal, although it is acknowledged that these may be too low. The use of variance may well be unsuitable in this concept, but we persist with it.

Problems also arise from the measurement of density, as flowering clones/m². Maximum readings from both populations are in the order of 13, but in many parts of the populations it can be as low as 0.5 to 1. However, as only approximately half the plants are available for crossing, due to the flower dimorphy, actual figures are half this. We do not at present have data which will give mean densities throughout the population, and neighbourhood sizes are estimated on the basis

Table 4. Estimates of neighbourhood size in *Primula veris*

Population/ year	Variance pollen travel (m) s^2_p	Variance seed travel (m) s^2_s	NS clones/m² 0.25	1	6.5
Dam 1, 1974	1.82	0.67	3.9	15.6	101.4
Whittle Dene Dam 1, 1975	3.37	0.67	6.3	25.4	165
Whittle Dene Dam 1, 1976	13.07	0.67	21.7	86.4	562
Park Shield 1976	1.75	1.77	5.6	22.1	144.1
Whittle Dene Dam 1, 1976 hybrid technique	5.45	0.31	9.0	35.4	233.4
Whittle Dene Dam 1, 1976 hybrid technique	0.73	0.10	1.2	5.1	33.6

of both extremes. It must be emphasised that these estimates of pollen travel will be underestimated at short distances, thus increasing the density dependent effect on travel and thus on neighbourhood size. Thus, although figures as varied as 1.2 to 562 are obtained for estimations of neighbourhood size (Table 4), much of this variability is due to differences in estimations of density. All in all, it seems reasonable to assume neighbourhood size between 5 and 200 in the populations under study, with an average in the order of 30. Nevertheless, even if density estimations are completely ignored, it is clear (Fig. 1) that isolated individuals or populations at low density do receive fewer insect visits, and thus neighbourhoods are highly density dependent.

DISCUSSION

There is very little data on the neighbourhood size of other plant species. It is likely that wind-pollinated species will show much larger values, and inbreeding annuals will show smaller values in some instances, although many inbreeding annuals tend to occur in dense stands. The only information available comes from an entomophilous outbreeder, like *Primula veris*. In *Phlox pilosa* (Levin & Kerster, 1968) neighbourhoods between 75 and 282 plants were estimated, not dissimilar to present results. Figures from animals tend to be much larger; thus Lamotte (1951) finds values of 5625 to 8400 in 2-dimensional populations, and 236 to 1050 in linear populations of *Cepaea nemoralis*, and Dobzhansky and Wright estimate 500–1000 in *Drosophila pseudobscura*. Plants, which show less motility, clonality, and relatively circumspect populations, will tend to show much lower neighbour-hood sizes, and figures for densely populated entomophilous outbreeders with low clonality, such as *Phlox pilosa* or *Primula veris*, are likely to be relatively high. Nevertheless, there are indications that at low density, neighbourhoods may drop to as low as ten individuals in *P. veris*.

Of the variables involved in the present estimations, seed travel is likely to be the most reliable. Estimations are low, probably genuinely so, and are unlikely to seriously affect final estimations. Estimations of variances of pollen travel are variable, by a factor of 15: this may reflect in part genuine differences between seasons and localities. However, underestimations of variance at high density and overestimations at low density inherent in the techniques are likely to render true variances smaller, and estimated variances may be on the high side. Unfortunately, the techniques used inevitably require different densities of plants, and thus it is not possible to make comparable estimations of travel at different densities. Undoubtedly, density will greatly affect true neighbourhood size, and this is likely to be, directly and indirectly, the most important variable. Despite arguments by Levin (1978), the much lower frequency of insect visits in areas of low plant density are likely to make neighbourhood calculations highly density-dependent and will not be compensated for by greater insect travel. The present data (Fig. 1) seems to clearly bear this out. As a result, in a species such as *P. veris*, sparsely populated populations of considerable size may have small neighbourhoods, as low as 10, or less.

It seems inevitable that estimations of neighbourhoods should be proble-matical, not only because of the inherent difficulty of measuring dispersal para-meters, but because they, and density parameters are so variable, even within a population. Also, although neighbourhoods may approximate to the number of individuals influencing the heterozygosity of their offspring, dormant seeds and non-flowering individuals within the neighbourhood, and continuously over-lapping neighbourhoods, will make the effective population number much higher. These considerations and the doubt of the validity of variance estimates in leptokurtic dispersal distributions, raise a serious question as to the value of neighbourhood size estimations. Although the present study has shown that it is possible to obtain data concerning pollen and seed dispersal, and density, it is doubtful whether these have much value except in small, circumspect populations with even density, in which the neighbourhood is chiefly governed by the limits of the population, which was far from the case in the populations studied. If these conditions are met, comparison of departures of genotype frequencies from panmictic expectations with population size estimations, may be of real interest.

Unfortunately, estimations of effective number (Kimura & Crow, 1963), based on the departure of frequencies of alleles and of heterozygotes from panmictic levels, involve circularity when genetic effects in small populations are under consideration. Nevertheless, these have shown effective population numbers in at least one plant, *Linanthus parryae*, to be well below 30, and thus liable to consider-able inbreeding effects (Wright, 1943). Doubtless, this is a frequent situation in plants, and although Ford (1964) may well be right in suggesting that genetic drift is unimportant in most animal populations, which tend to be much larger, it may well be of great importance in plants. Until we are able to compare isolated neighbourhoods with frequencies of alleles and heterozygotes, however, even circumstantial evidence for this will continue to be lacking.

REFERENCES

BATEMAN, A. J., 1946. Contamination of seed crops. I. Insect pollination. *Journal of Genetics, 48:* 257.
BATEMAN, A. J., 1947. Contamination of seed crops. III. Relation with isolation distance. *Heredity, 1:* 303–336.
BODMER, W. F., 1960. The genetics of homostyly in populations of *Primula vulgaris. Philosophical Transactions of the Royal Society* (B), *242:* 517–549.
CROSBY, J. L., 1949. Selection of an unfavourable gene-complex. *Evolution, 3:* 212–230.
DOBZHANSKY, Th. & WRIGHT, S., 1943. Genetics of natural populations. X. Dispersal rates in *Drosophila pseudoobscura. Genetics, 28:* 304–340.
DOWRICK, V. P. J., 1956. Heterostyly and homostyly in *Primula obconica. Heredity, 10:* 219–236.
FORD, E. B., 1964. *Ecological Genetics.* London: Methuen.
FREE, J. B., 1962. The effect of distance from pollinizer varieties on the fruit set on trees in plum and apple orchards. *Journal of Horticultural Science, 37:* 262–271.
FREE, J. B., 1970. *Insect Pollination of Crops.* London: Academic Press.
HARBERD, D. J., 1961. Observations of the population structure and longevity of *Festuca rubra. New Phytologist, 60:* 184–206.
HARBERD, D. J., 1962. Some observations on natural clones in *Festuca ovina. New Phytologist, 61:* 85–100.
KERSTER, H. W., 1964. Neighbourhood size in the rusty lizard, *Sceloporus olivaceus. Evolution, 18:* 445–457.
KIMURA, M. & CROW, J. F., 1963. The measurement of effective population number. *Evolution, 17:* 279–288.
LAMOTTE, M., 1951. Récherches sur la structure génétypique des populations naturelles de *Cepaea nemoralis. Bulletin biologique de la France et de la Belgique, 35:* 1–239.
LEES, D. R., 1971. Frequencies of pin and thrum plants in a wild population of the Cowslip, *Primula veris. Watsonia, 8:* 289–291.
LEVIN, D. A., 1972. Pollen exchange as a function of species proximity in *Phlox. Evolution, 26:* 251–258.
LEVIN, D. A., 1978. Pollinator behaviour and the breeding structure of plant populations. In A. J. Richards (Ed.), *The Pollination of Flowers by Insects.* London: Academic Press.
LEVIN, D. A. & KERSTER, H. W., 1968. Local gene dispersal in *Phlox. Evolution, 22:* 130–139.
LEVIN, D. A. & KERSTER, H. W., 1969. Density dependent gene dispersal in *Liatris. American Naturalist, 103:* 61–74.

LEWIS, D., 1954. Comparative incompatibility in Angiosperms and Fungi. *Advances in Genetics, 6:* 235–281.

PROCTOR, M. C. F. & YEO, P. F., 1973. *The Pollination of Flowers.* London: Collins.

STEPHENS, S. G. & FINKER, M. D., 1953. Natural crossing in cotton. *Economic Botany, 7:* 257–269.

VALENTINE, D. H., 1975. *Primula.* In C. A. Stace (Ed.), *Hybridisation and the Flora of the British Isles.* London: Academic Press.

WRIGHT, S., 1938. Size of population and breeding structure in relation to evolution. *Science, 87:* 430–431.

WRIGHT, S., 1943. Isolation by distance. *Genetics, 28:* 114–138.

WRIGHT, S., 1946. Isolation by distance under diverse systems of mating. *Genetics, 31:* 39–59.

WRIGHT, S., 1948. On the roles of directed and random changes in gene frequency in the genetics of populations. *Evolution, 2:* 279–294.

The role of preferential and assortative pollination in the maintenance of flower colour polymorphisms

Q. O. N. KAY

Department of Botany and Microbiology, University College of Swansea, Swansea, Wales

Flower-colour polymorphisms occur in wild populations of many angiosperms, but little is known of the selective factors that may maintain the polymorphisms, and there have been few observations of the behaviour of the pollinators of polymorphic species. There is good evidence that some lepidopteran and dipteran pollinators show strong preferences for particular colour morphs of polymorphic species, for example in *Leavenworthia crassa* and *Lantana camara*. Recent work on *Raphanus raphanistrum*, which has a conspicuous white/yellow (bee-purple) flower-colour polymorphism has shown that *Pieris* and *Eristalis* spp. show a very strong preference for yellow flowers in polymorphic populations, while some *Bombus* spp. show a preference for white flowers in the same populations. Flower-colour polymorphisms of this type may be maintained by mechanisms which depend on the behaviour of discriminatory pollinators in relation to the relative frequencies of the flower-colour morphs and of competing flowers of other species.

KEY WORDS:—flower colour polymorphism—assortative pollination—*Bombus*—*Eristalis*—*Pieris*—*Raphanus*.

CONTENTS

NATURE AND OCCURRENCE OF FLOWER COLOUR POLYMORPHISM

Many insect-pollinated plant species show variation in flower colour. This often depends only on the microenvironment or age of the flower, and is purely phenotypic with no genetic basis. But in a large number there is also some degree of genetic variation. The commonest type of flower colour variation is continuous, covers a relatively small colour range, and probably depends on small differences in the distribution and concentration of the floral pigments, or in the structure of the petals and other floral parts. Although this type of continuous variation is easily observed, it is very difficult to quantify and to investigate patterns of intra- and interpopulation variation, and possible discrimination by pollinators. A related type of variation occurs in some autogamous (e.g. *Euphrasia* spp.) and apomictic (e.g. *Hieracium* spp.) groups, in which different biotypes often show minor but constant differences in flower colour; while these differences may be of taxonomic

175

value, they are probably of little if any selective importance in most cases, although they may be of some importance in the autogamous groups if there is any degree of outcrossing.

A second type of flower colour variation is less frequent but is of great interest. This is genetic polymorphism in flower colour, the situation in which two or more genetically different phenotypes with different and clearly distinguishable flower colours (flower colour morphs) coexist at higher frequencies than those maintained by mutation, in the same interbreeding population (gamodeme) of a species. A fairly large number of common and widespread species show this type of variation in part at least of their British ranges; for example *Anagallis arvensis* (red, pink and blue), *Cirsium palustre* (purple, pale purple and white), *Erica tetralix* (purplish pink and white), *Galeopsis tetrahit* (purple, pink and white, *Orchis mascula* (purple, pale purple, white, etc.), *Polygala serpyllifolia* and *P. vulgaris* (blue, red and white), *Primula vulgaris* (yellow, pink and white), *Raphanus raphanistrum* (yellow and white), *Vicia sativa* sensu lato (purple and pink) and *Viola lutea* and *V. tricolor* sensu lato (blue-violet, yellow, etc.). Many cases of flower colour polymorphism of this type are probably balanced polymorphisms, but very little is known about the factors that cause and maintain them. In some cases, flower colour variation which may or may not involve genetic polymorphism appears to result from inter-specific hybridisation and is found only in hybrid or introgressed populations, for example in *Medicago* × *varia* (*M. falcata* × *M. sativa*), *Raphanus maritimus* (Harberd & Kay, 1975) and *Silene dioica* × *S. alba* (Baker, 1948).

In a number of species, flower colour morphs that normally appear only as rare mutants occur at relatively high frequencies in a few scattered populations. Genetic drift may be involved here, with the polymorphic populations being founded by small groups of plants that happen to include the rare morph. White-flowered morphs often show this pattern of very local occurrence at relatively high frequencies, for example in *Digitalis purpurea, Ononis repens, Scilla verna* and *Succisa pratensis*. In these cases the rarer form may be at a selective disadvantage and may eventually revert to its normal frequency. Another situation in which flower colour forms that are normally at a selective disadvantage may be temporarily established at high frequencies in spontaneous or subspontaneous populations occurs when these populations are descended from cultivated populations of crop or ornamental species in which different flower colour forms have been artificially maintained or brought together. Ruderal populations of *Antirrhinum majus, Kentranthus ruber* (Plate 21), and *Papaver somniferum* often show this type of variation, and there are many other examples. In some cases that appear to be of this type, the wild species from which the cultivars were derived may have had a balanced flower colour polymorphism which is re-established in ruderal or naturalised populations. It is, for example, not known whether the yellow/pink flower colour polymorphism of *Primula vulgaris* populations in western Britain is a primary balanced polymorphism that arose in native populations, or a secondary balanced polymorphism derived by introgression from pink-flowered garden plants (of Mediterranean origin) into native yellow-flowered populations, or again a transient polymorphism depending on the presence of pink-flowered plants temporarily naturalised from garden populations. *P. vulgaris* is polymorphic for flower colour in native populations in southern and south-eastern Europe.

It is perhaps significant that some of the most conspicuous examples of apparently balanced flower colour polymorphism occur in outbreeding, locally abundant ruderal species that are adapted to a wide range of pollinators, for example *Chrysanthemum coronarium* (yellow/white) and *Raphanus raphanistrum*. Many of the species that show flower colour polymorphism are, like *C. coronarium* and *R. raphanistrum*, annuals or biennials that grow in open or disturbed habitats and show great fluctuations in population size from year to year; other examples of this type are *Linanthus parryae* (Epling & Dobzhansky, 1942; Epling, Lewis & Ball, 1960), *Leavenworthia* spp. (Lloyd, 1969) and *Cirsium palustre* (Mogford, 1974a). In all these cases the flower colour morphs are well differentiated, and morph ratios may change widely over relatively small distances, although the morph ratio often appears to be stable from year to year in a local population. Epling *et al.* (1960) found that the frequencies of the blue and white petal colour morphs of *Linanthus parryae* were remarkably stable in local populations; the overall pattern of variation was irregularly clinal. Lloyd observed that the relative frequencies of the petal colour morphs of *Leavenworthia crassa* showed little variation within each old-established glade population, but showed great variation in secondary populations on cultivated land, where a population could change from predominantly yellow to predominantly yellow-centred over a distance of less than 20 m. He suggested that genetic drift might have caused the heterogeneity of these secondary populations. I have observed a similar pattern of variation in polymorphic *Chrysanthemum coronarium* in Greece, where presumably old-established populations on long-cultivated land are homogeneous, with little variation in the relative frequencies of the yellow (insect yellow) and white (insect blue-green) flower colour morphs within each group of fields. However, populations growing in temporarily available roadside and ruderal habitats are heterogeneous, with great changes in morph frequency over distances of a few metres. Mogford's maps of flower colour morph frequency in *Cirsium palustre* in this volume show that the centres of populations with widely different morph ratios may be separated by only 150–200 m, indicating sharp local changes in morph frequencies. The relative frequencies of the flower colour morphs of *C. palustre* also show geographic and altitudinal clinal variation on a large scale.

In *Raphanus raphanistrum*, which is a weed of cultivated land, I have found that the relative frequencies of the yellow (insect purple) and white (insect white) petal colour morphs are normally uniform and stable within each farm or group of farms where the plant is abundant. Large changes in relative frequencies of the petal colour morphs usually occur clinally, over distances of 20–50 km in Britain, in populations growing in arable land. Roadside and ruderal populations of *R. raphanistrum* do however frequently show heterogeneity and differences from the prevailing local morph ratio, which could result from genetic drift or founder effect.

POSSIBLE CAUSES OF FLOWER COLOUR POLYMORPHISM

Flower colour polymorphisms are of particular evolutionary interest because pollinators may discriminate between the flower colour morphs, and thus affect pollination, seed production, gene flow and recombination in the different morphs, and in polymorphic or monomorphic populations. There are several possible ways in which a balanced flower colour polymorphism could be maintained by selective pollination.

Disassortative pollination, with pollinators alternating between the different morphs would (a) maintain a stable polymorphism with an excess of heterozygotes. There is as yet no evidence for such a phenomenon.

Assortative pollination, with a pollinator preferring one or only some of the morphs during each series of visits, would not maintain a stable polymorphism unless (b) a heterozygote was preferred to both homozygotes; or (c) the heterozygote, and a homozygote that was at a selective disadvantage in other respects (it could, for example, be more conspicuous to seed-eating animals) were preferred to the other homozygote; or (d) there was frequency-dependence with a flower colour phenotype being preferentially visited when its frequency in the population fell below a certain level.

If two or more pollinators were involved, and at least one pollinator preferred one or more of the flower colour morphs, while other pollinators did not discriminate, a stable polymorphism could be maintained by a simple form of frequency-dependent advantage (e) in which the preferred morph was at a small selective disadvantage in other respects, and the visits by the preferential pollinator increased in relative frequency as the frequency of the preferred morph fell in the population

If different pollinators preferred different flower colour morphs, a fluctuating polymorphism could be maintained if (f) the relative numbers of pollinators with different preferences changed from year to year and season to season.

Similarly, but in space rather than in time, a mosaic polymorphism would be maintained if (g) the relative numbers of pollinators with different preferences differed in adjacent habitats within the range of the polymorphic species. This type of unstable polymorphism could also be maintained by a single pollinator species if its preferences differed in different seasons and different habitats; its preference for a particular morph might, for example, depend upon the relative abundance of another species with similar flowers.

In practice these pollinator-mediated mechanisms are likely to be combined in various ways, with a variety of mechanisms interacting to maintain each flower colour polymorphism. The possibility that the same flower colour polymorphism may be maintained by different pollinator-mediated mechanisms in different parts of the range of the polymorphic species should also be considered, as should the possibility that some flower colour polymorphisms may be maintained partly or completely by mechanisms that do not involve pollinators.

POLLINATOR DISCRIMINATION

Most of the more active insect pollinators probably or certainly have trichromatic colour vision extending into the near ultraviolet (Daumer, 1956; Mazokhin-Porshnyakov, 1962; Struwe, 1972a, b; Bishop & Chung, 1972). The common practice of describing flower colour variation only in terms of the colours visible to the human eye is therefore unsatisfactory. Few investigators have given any information about the ultraviolet components of the flower colour forms that they have studied. The best procedure is probably to obtain complete reflectance spectra for the petal colours that are visible to the pollinators, as was done by Kauffeld & Sorensen (1971). Many flower colour polymorphisms probably involve colours with

UV components which differ between visible morphs; additional UV morphs with differences that are not visible to the human eye may also exist in some cases, and the frequency of flower colour polymorphisms that are restricted to UV wavelengths is unknown.

The majority of the rather small number of investigations of pollinator behaviour in relation to flower colour variation within a species have been made using mixed populations of cultivars which may differ from one another in many respects apart from flower colour. Bateman (1951), for example, observed discrimination by *Apis mellifera* and *Bombus* spp. between cultivars with different flower colours in both cabbage (*Brassica oleracea*) and swede (*B. napus*). Free & Williams (1973) and Faulkner (1976) similarly observed discrimination by *Apis mellifera* between cultivars of brussels sprout which differed in flower colour, height and other characters. Kauffeld & Sorensen (1971) found that *Apis mellifera* discriminated visually between cultivars of *Medicago sativa* which differed in flower colour. There was a close correlation between the nectar yield of each cultivar and the order of preference shown by *A. mellifera* in the field, where purple-flowered cultivars were preferred. In a separate experiment in which *M. sativa* flowers were exposed on uniform plastic discs, they found that individuals of *A. mellifera* which had been foraging on white and yellow sweet clover preferred white and yellow cultivars of *M. sativa*. The 25 cultivars that they studied differed in several respects in addition to flower colour and nectar yield. Pedersen & Todd (1949) had earlier reported differences in attractiveness to *Apis mellifera* among 14 different cultivars of *M. sativa*. Clement (1965) had however reported that in a mixture of two isogenic lines of *M. sativa* differing only in flower colour, 3 out of 17 marked individuals of *Apis mellifera* preferred purple flowers, while one preferred white flowers.

Goplen & Brandt (1975) found that yellow-flowered morphs of the predominantly purple-flowered *M. sativa* cv. 'Ladak' were tripped very much less frequently and produced about half as much seed as purple-flowered morphs in the same population, when pollinated by leaf-cutter bees (*Megachile pacifica*). Pedersen (1967) had previously reported that pollen-collecting leaf-cutter bees preferred coloured to white flowers in *M. sativa*.

Kugler (1955) reported that *Bombus* spp. usually did not discriminate between red-flowered and white-flowered cultivars of *Lathyrus odoratus*, but could be induced to confine their visits to the red form if syrup was placed in its flowers, or if the nectar of the white flowers was diluted. Leleji (1973) compared the attractiveness to bees of a white-flowered cultivar of the cowpea (*Vigna sinensis*) with that of a purple-flowered but otherwise similar cultivar, in an experimental mixed population in northern Nigeria. He found that bumble bees, which were apparently seeking nectar, visited purple flowers more often than white (3:1) whereas honey bees, mainly seeking pollen, visited purple flowers less often than white (1:2). He did not investigate possible differences in nectar yield and scent between the cultivars.

Erickson (1975a, b) stated that the cultivars of soybean (*Glycine max*) that are most attractive to honey bees have white flowers; he investigated several hundred cultivars, which varied from purple to white in flower colour, and also showed a wide range of variation in many other characters.

Thus, it has been shown that in *Medicago sativa* some pollinators (*Apis mellifera* and *Megachile pacifica*) discriminate between flower colour forms which probably do not differ in other characters. But in other instances of discrimination between flower colour forms, both in *M. sativa* and in other cultivated plants, discrimination by pollinators may be caused by other differences between the colour forms, although the colour forms are certainly recognised visually in some cases.

There have been very few studies of pollinator behaviour in natural or experimental populations of wild species showing flower colour polymorphism. It does however appear that in all the cases in which substantial numbers of observations have been made, significant and in some instances very strong preferences for particular flower colour morphs are shown by at least some of the pollinators. Dronamraju (1960) observed significant numbers of visits by five species of butterfly to a small experimental population (three bushes of each flower colour) of pink-flowered and orange-flowered *Lantana camara* in Calcutta. He found that *Precis almana* (Nymphalidae) showed a strong preference for orange flowers (218 visits to orange flowers: 13 visits to pink flowers), whereas *Papilio demoleus* (Papilionidae), *Catopsilia pyranthe* (Pieridae) and *Baoris mathias* (Hesperiidae) discriminated strongly against orange flowers (42:98, 40:603 and 1:108 respectively). *Danaus chrysippus* (Danaidae) did not discriminate (142:158). He stated that both colour forms were common in ruderal habitats in Calcutta, but gave no information about the genetic interrelationships of the forms. Fritz Müller had observed in the late nineteenth century that different species of butterfly showed preferences for different colours of flowers in *Lantana camara* (Wallace, 1889: 317).

Lloyd (1969) observed visits by pollinating insects in two adjacent subpopulations of a self-incompatible race of *Leavenworthia crassa* in which three flower colour morphs occurred. He found that *Apis mellifera* did not discriminate between the two commoner morphs (yellow-centred and yellow) but did show a small preference for the rarest morph (eye-intermediate), making rather more than twice as many visits to it than would be expected, in the most extreme case (10.8% of visits to eye-intermediate in a population in which its frequency was 4.1%, $\chi^2 = 10.81$). Solitary bees, which may be the major natural pollinators of *L. crassa*, did not discriminate between yellow and yellow-centred but made a small though not significant excess of visits to eye-intermediate. In contrast, *Bombylius major* (Diptera, Bombyliidae), which was also an important visitor, showed a very strong preference for the commonest morph (yellow-centred); *B. major* made 94.1% of its visits to yellow-centred in a population in which the frequency of yellow-centred was 60.3% (185 visits observed, $\chi^2 = 48.3$), 100% of its visits to yellow-centred in a population in which the frequency of yellow-centred was 74.7% (122 visits observed $\chi^2 = 34.1$) and 85.3% of its visits to yellow-centred in a population in which its frequency was 41.8% (245 visits observed, $\chi^2 = 150.0$). The butterfly *Anthocharis genutiae* (Pieridae) did not discriminate between the morphs. Yellow was recessive to yellow-centred. Lloyd did not investigate the ultraviolet components of the petal colours (pers. comm.).

Mogford (1974b) observed visits by *Bombus* spp. and *Apis mellifera* to polymorphic populations of *Cirsium palustre* in which purple, pale purple (intermediate), purplish white and white flower colour morphs occurred at various frequencies. He found that the white or purplish white morphs were preferred to some extent by some, but not all, *Bombus* spp.; the preferences were not very strong, and bees of the same species of *Bombus* showed significant preferences in some populations

but not in others. For example, the greatest discrimination observed in any population was shown by *B. lapponicus* at Forest Lodge, where it made 88.4% of its visits
to the white morph in a population with only 64.0% of the white morph, but at
Cantref, a few kilometres away, *B. lapponicus* did not discriminate significantly in a
population with 25.5% of the white morph. In both cases *B. lapponicus* formed only
a small proportion of the total number of visitors (1.0% and 1.3% respectively),
and it is possible that individual bees may have differed in their colour preferences,
perhaps majoring (Heinrich, 1976) on different morphs.

Levin and his co-workers (Levin, 1969, 1972a, b; Levin & Kerster, 1970; Levin &
Schaal, 1970) have made an interesting series of studies of natural and experimental populations of *Phlox* spp. in which the relative attractiveness of different
flower colour forms to pollinators were assessed indirectly by comparing pollen
load on the stigma, and seed set. They found clear differences in attractiveness
between different flower colour morphs or other flower colour forms (different
cultivars, for example). The pollinators of the *Phlox* spp. that they studied are
mainly Lepidoptera, and Levin concluded that butterflies may act as strong selective
agents upon flower colour. The differences in overall attractiveness were, however,
relatively small; in artificial populations of mixed cultivars of *Phlox drummondii*,
for example, the difference in attractiveness between the least attractive colour form
(coral) and the most attractive colour form (lavender) was 1:1.49. Levin's
demonstration of low frequency disadvantage in artificial mixtures of *Phlox
drummondii* cvs. 'Nana compacta' and 'Twinkle' was particularly interesting, but as
these cultivars differ in several respects the effect may be due to factors other than
their flower colour differences. Levin & Kerster (1973) have found that there is
strong assortative pollination for height by *Apis mellifera* in both natural and
artificial populations of *Lythrum salicaria*; similar assortative pollination is known
to occur in mixtures of brussels sprout cultivars that differ in height (Faulkner,
1976). The indirect techniques that were used by Levin and his co-workers in their
studies of *Phlox* spp. would not detect assortative pollination in most cases, and
further studies involving direct observation of pollinator behaviour and progeny
analyses are needed.

ASSORTATIVE POLLINATION IN POLYMORPHIC *RAPHANUS RAPHANISTRUM*

Raphanus raphanistrum (Cruciferae), an annual weed of cultivated land, is a
particularly striking and interesting case of flower colour polymorphism. Sampson
(1964, 1967) reported that the populations of *R. raphanistrum* that he studied were
strongly self-incompatible, with between 15 and 82 self-incompatibility alleles (95%
confidence limits). I have found consistent and strong self-incompatibility in small
samples of plants from 20 widely separated European populations (from Sweden,
Germany, England, Scotland, Wales, France, Portugal and Greece) and wild
populations appear to be uniformly self-incompatible. It is polymorphic for two
conspicuous corolla characters in Britain; petal colour, which is either white or
yellow, and petal veining, with dark anthocyanin veins being either present or
absent. The colour polymorphism is not linked to the veining polymorphism,
although they show broadly parallel clines in the British Isles. The petal colour
morphs differ only in a single allele, and white is dominant to yellow. There is some
variation in the intensity of the yellow colour. Reflectance spectra of unveined

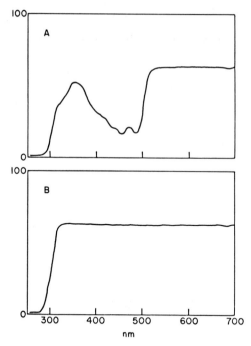

Figure 1. Reflectance spectra of petals of unveined yellow-flowered (A) and white-flowered (B) morphs of *Raphanus raphanistrum* from a population at Sketty, Swansea. The spectra were obtained using a Unicam SP 8000 recording spectrophotometer fitted with an SP 890 diffuse reflectance unit.

yellow and white petals are shown in Fig. 1. The petals of the yellow morph have a typical carotenoid reflectance spectrum with a peak of reflectance in the near ultraviolet, and are thus insect-purple. The petals of the white morph show high and uniform reflectance throughout the visible and near ultraviolet spectrum and are thus insect-white, a surprisingly rare petal colour, which I have also observed in white-flowered *Brassica oleracea*, and Horovitz & Cohen (1972) have observed in the Mediterranean crucifers *Leptaelum filifolium* and *Carrichtera annua*. The petals of most white-flowered species in the Cruciferae, like the petals of the great majority of white-flowered species in other families, absorb ultraviolet light of the wavelengths visible to insects and are thus insect blue-green. Daumer (1958) did not report finding any insect white flowers in a survey of 204 species. The exceptionally high reflectance of the petals of white *R. raphanistrum* flowers makes them particularly conspicuous in dim light.

Mixed populations of white-flowered and yellow-flowered morphs of *R. raphanistrum* are common in southern and central Britain and also occur widely in north-western and central Europe, and in other areas of the world to which *R. raphanistrum* has spread as a weed, for example the United States. In the Mediterranean region most populations are monomorphic, with pale yellow veined petals, while in the oceanic margin of western Europe (Brittany, Cornwall, western Wales, Ireland and much of Scotland) and probably also in parts of Scandinavia and northern Germany, most populations are monomorphic, with

Figure 2. Frequencies of yellow-flowered morphs (black sectors) and white-flowered morphs (white sectors) in population samples of *Raphanus raphanistrum* observed in Britain between 1965 and 1975.

bright yellow unveined petals. The distribution of monomorphic and poly-morphic populations is not yet fully known. The unveined yellow morphs of north-western Europe are frequently confused with other yellow-flowered annual Cruciferae, for example *Sinapis arvensis* and *Brassica rapa*; this has led to consider-able under-recording of the yellow morph, and of *R. raphanistrum* in the areas in which the yellow morph forms monomorphic populations. This can be seen in the distribution maps given for the species (Perring & Walters, 1962) and for the morphs (Perring, 1968). The pale yellow flowers of typical Mediterranean populations of *R. raphanistrum* appear almost white to the human eye in bright sunlight, and have frequently been described as white by visiting botanists. In the shade they can be seen to be yellow, and they are in fact very similar in colour intensity to the yellow-flowered morphs of polymorphic populations in southern Britain.

The frequencies of different flower colour morphs in population that I have observed in Britain are shown in Fig. 2. Most populations in southern and eastern England consist predominantly of white-flowered plants, with yellow-flowered plants present at a low frequency (often less than 2%) in nearly all populations.

Yellow-flowered plants occur at higher frequencies in a few populations in this area. In the northern and western English Midlands and in south-eastern Wales, most populations are mixed, with fairly high frequencies of both flower colour morphs. In northern England, western Wales, and Scotland apart from the Central Lowlands, most populations consist solely of yellow-flowered plants; mixed populations are reported to occur in the Scottish Lowlands. Records in regional Floras and other sources (e.g. Johnston, 1829; Ray, 1724; Smith, 1804) suggest that the distribution and relative frequencies of the morphs have been fairly stable at least since the early nineteenth century, although there have been some suggestions that the conspicuous white-flowered morph may have increased to some extent.

I have observed the behaviour of pollinating insects in six polymorphic populations of *R. raphanistrum*, with frequencies of the yellow-flowered morph ranging from 0.8% to 68.3%, in central England and southern Wales during 1975. Bumble bees (*Bombus* spp.) and sometimes honey bees (*Apis mellifera*) are the chief visitors to *R. raphanistrum* in this region, but in all populations cabbage white butterflies (*Pieris* spp.), especially *P. rapae* (Lepidoptera, Pieridae) were also important, often making 10% to 15% of the total number of visits to *R. raphanistrum*. *Eristalis* spp., *Syrphus* spp. and other hoverflies (Diptera, Syrphidae) were common but minor visitors, making less than 5% of the visits to *R. raphanistrum*, and some nymphalid butterflies (*Aglais urticae*, *Vanessa cardui*) and noctuid moths (chiefly *Plusia gamma*) made smaller numbers of visits in some populations. The relative pollinating efficencies of the different insects were not investigated quantitatively, but all made legitimate visits and appeared to carry and transfer pollen grains in sufficient numbers to fertilise all the ovules in at least the first few flowers that they visited after moving from one plant to another (less than 12 ovules normally function in each flower of *R. raphanistrum*).

Kendall & Solomon (1973) estimated the quantities of pollen carried on the bodies (excluding pollen loads on bees) of more than 70 species of insect that were visiting the flowers of apple trees. The mean numbers of pollen grains of all types that they estimated to be present on the bodies of their samples of individuals of the species that I have observed to visit *R. raphanistrum* were as follows: *Bombus pascuorum* 4650, *B. lapidarius* 11050, *B. terrestris* 9100 to 1900, *Apis mellifera* 5650. *Eristalis arbustorum* 2350, *E. tenax* 3550, *Helophilus pendulus* 2100, *Syrphus* spp. 350, *Aglais urticae* 1550, *Pieris napi* (two individuals) 8. I have however observed much greater numbers of pollen grains on *Pieris* spp. in the field, adhering mainly to the head and proboscis. Levin & Berube (1972) describe observations on the pollination of *Phlox* by the butterfly *Colias eurytheme* (Pieridae) in which they found that the proboscis of *C. eurytheme* picked up means of 1950 grains from *P. pilosa* and 1050 grains from *P. glaberrima;* single pollinations of *P. pilosa* delivered means of 365 grains, and of *P. glaberrima* 101 grains. Similar quantitative observations on the pollination of *R. raphanistrum* by *Pieris* spp. would be of great interest. Although my field observations indicated that *Pieris* spp. carried fewer pollen grains of *R. raphanistrum* than did bees, *Pieris* spp. had similar or greater rates of working (Table 3), normally flew much greater distances between the plants that they visited, and visited only one or a few flowers on each individual plant. It thus appeared possible that *Pieris* spp. could be more efficient agents of gene-flow and fertilisation than bees. These tended to fly only short distances between closely

Table 1. Frequency of pollinator visits to the yellow-flowered morph of *Raphanus raphanistrum* in five polymorphic populations

| Site | Frequency of yellow morphs in population (%) | % of total visits made to yellow flowers | | | | | |
| | | *Pieris rapae* | | | P. | *Eristalis* | |
		Males	Females	*P. napi*	brassicae	arbustorum	*E. tenax*
Sketty B	60.8	94.5**	99.7**	99.1**	100**	89.0**	87.7**
	(418)	(292)	(307)	(111)	(107)	(556)	(235)
Brownhills	42.3	89.8**	81.1**	n.d.	n.d.	66.0*	70.7**
	(130)	(303)	(169)			(47)	(75)
Singleton	27.4	92.3**	89.3**	93.7**	n.d.	96.0**	n.d.
	(73)	(1769)	(131)	(142)		(569)	
Beenham	7.3	27.4**	25.8**	n.d.	n.d.	n.d.	40.8**
	(303)	(446)	(190)				(49)
Landimore	0.8	4.8**	0.0	0.0	n.d.	n.d.	n.d.
	(240)	(268)	(233)	(470)			

The size of the sample is shown in parentheses. Significant differences from the population morph frequencies are indicated by single ($P<0.01$) or double ($P<0.001$) asterisks. The grid references of the sites are as follows: Sketty, SS 623921; Singleton, SS 629917; Brownhills, SK 068052; Beenham, SU 592690; Landimore, SS 473926.

Table 2. Frequency of visits to yellow-flowered *R. raphanistrum* by bumblebees and and honeybees working mixed populations

| Site | % of yellow morphs in population | % of total visits made to yellow flowers | | | |
		Bombus pascuorum	*Bombus lapidarius*	*Bombus terrestris*	*Apis mellifera*
Sketty A	68.3	59.81^W	—	—	68.21
		(938)			(453)
Sketty B	60.8	52.89^W	62.51	64.48^y	71.59^Y
		(2265)	(835)	(1329)	(1232)
Brownhills	42.3	—	—	35.35^W	—
Singleton	27.4	13.62^W	—	35.10^Y	38.55^Y
		(2598)		(3892)	(1284)
Beenham	7.3	4.27^W	4.25^W	—	8.71
		(772)	(729)		(574)
Landimore	20.0	20.75	11.08^W	—	—
	(0.8)	(805)	(325)		

The size of each sample is shown in parentheses. Significant differences from the population morph frequencies are indicated thus: y, Y, *Y*, excess of visits to yellow flowers, and w, W, *W*, excess of visits to white flowers, significant at 0.05, 0.01, and 0.001 levels respectively.

adjacent plants, and visited large numbers of flowers in succession on each individual plant, perhaps carrying out more ineffective self-pollination (geitonogamy) than compatible cross-pollination.

The relative numbers of visits made to the white-flowered and yellow-flowered morphs by the different species of visitors are shown in Tables 1, 2 and 3. Most visitors discriminated between the morphs to a significant extent in some or all of the mixed populations, and in some cases their preferences were extremely strong. The chief visitors, *Bombus* spp. and *Apis mellifera*, showed significant but less strong preferences (Table 2). *Bombus pascuorum* preferred the white-flowered morphs in all the mixed populations (the Landimore population had a very low overall frequency of yellows, and the behaviour of *Bombus* spp. was observed there in a small area in which the proportion of yellow-flowered morphs was artificially increased by removing white-flowered plants; these observations are probably not comparable with those made in other populations). *B. lapidarius* showed some preference for the white-flowered morph in one of the two mixed populations in which its behaviour was observed. *B. terrestris* was inconsistent, showing a small but significant preference for yellow-flowered morphs in two populations and a preference for the white-flowered morph in another. *Apis mellifera* did not discriminate in two of the four mixed populations that were observed, but showed a significant preference for yellow-flowered morphs in the other two; in both of the

Table 3. Frequency of visits to yellow-flowered *R. raphanistrum* by pollinators working a mixed population at Singleton with 27.4% yellow-flowered plants

Species	% of total visits made to yellow flowers	Mean duration of visits (sec)
(Apidae)		
B. terrestris	35.10^Y (3892)	4.01
Bombus pascuorum	13.62^W (2598)	4.87
Apis mellifera	38.55^Y (1284)	6.80
(Syrphidae)		
Eristalis arbustorum	96.00^Y (569)	11.30
Helophilus cf. *pendulus*	93.55^Y (31)	—
Syrphus cf. *venustus*	33.33 (129)	8.57
Syrphus cf. *vitripennis*	28.32 (459)	19.02
(Lepidoptera)		
Pieris rapae males	92.31^Y (1769)	4.17
P. rapae females	89.31^Y (131)	6.46
P. napi	93.66^Y (142)	5.90

The size of each sample is shown in parentheses. Significant differences ($P<0.001$) from the population morph frequency are indicated thus: *W*, white flowers preferred; *Y*, yellow flowers preferred.

latter *B. pascuorum* was abundant and particularly active, preferring white-flowered morphs. In several cases, relatively few individuals of each bee species were observed, and it is possible that different individuals may show different preferences, perhaps majoring (Heinrich ,1976) on different morphs.

The most striking preferences were those shown by *Pieris* spp. and *Eristalis* spp. (Table 1). Both showed a strong or very strong preference for the yellow-flowered morph in four populations in which the frequencies of the yellow-flowered populations ranged from 7.3% to 60.8%. At the Sketty B site, for example, where the frequency of the yellow-flowered morph was 60.8%, 307 visits by female *P. rapae* to *R. raphanistrum* were observed; 306 of these visits were to yellow flowers and only one to a white flower. *P. napi* and *P. brassicae* showed an equally strong preference for the yellow-flowered morph at the same site. At this site, and at Brownhills and Beenham, *R. raphanistrum* was abundant, with yellow-flowered and white-flowered morphs closely intermingled in a mixed stand of field weeds. The Singleton population, which was the only artificial population, consisted of well-spaced rows of F_2 plants segregating at random for yellow or white flower colour. Although the frequency of yellow-flowered plants was only 27.3% in this population, the proportion of visits made to yellow flowers by *P. rapae* was higher than at the Brownhills site (where there were 42.3% yellows) and the proportion of visits made to yellow flowers by *Eristalis arbustorum* was higher than at any other site, perhaps as a result of the spatial separation of individual plants at Singleton. The absence of discrimination shown by *Syrphus* spp. contrasts with the strong discrimination shown by *Eristalis* spp. (Table 3).

The Landimore population differed from the others both in the low frequency of the yellow morph (0.8%) and in the abundant occurrence of *Sinapis arvensis*, a crucifer with yellow flowers which resemble those of *R. raphanistrum* and have a similar reflectance spectrum to that of yellow *R. raphanistrum* (Kay, 1976). *E. arbustorum*, *E. tenax* and other *Eristalis* spp. were abundant on *S. arvensis* flowers at this site, but did not visit *R. raphanistrum*. *P. rapae* and *P. napi* visited white-flowered *R. raphanistrum* and *S. arvensis* with approximately equal frequency. Only two cases of apparently preferential visits to yellow-flowered *R. raphanistrum* were observed at Landimore; a male *P. rapae* was involved in each case. Most male *P. rapae*, and all *P. napi* and female *P. rapae*, did not seek yellow-flowered *R. raphanistrum* there.

In all populations, the distances that were flown by individuals of the *Pieris* spp. between visited plants were commonly one to several metres, and flights of 15–40 m between successively visited plants were not unusual, with total distances of at least 150 m being covered during a series of visits. Recognition and location of yellow-flowered plants seemed to be entirely visual. Differences in behaviour between different individuals of *P. rapae* were observed at all sites, but were most obvious at the Beenham site, where less than half of the individuals of both sexes sought yellow-flowered *R. raphanistrum*. In the natural populations, the strength of the preference that *P. rapae* showed for the yellow morph clearly decreased with decreasing frequency of the yellow morph. *Pieris* spp. are known to show great changes in abundance from year to year, and such changes in the relative abundance of the preferential pollinators of *R. raphanistrum* are likely to affect the relative reproductive success of the flower colour morphs.

The larvae of *Pieris* spp. feed on a range of crucifers including *R. raphanistrum*, which is occasionally severely damaged by larvae of *P. brassicae*. It thus seems possible that the yellow-flowered morph of *R. raphanistrum* might, because of the greater attractiveness of its flowers to *Pieris* spp., suffer greater damage from the attentions of *Pieris* larvae than the white-flowered morph. I have, however, observed that female *P. rapae* do not appear to discriminate between the yellow and white flower colour morphs when they are engaged in egg-laying. At Singleton, where the frequency of yellow-flowered morphs was 27.4%, female *P. rapae* made 89.3% of an observed total of 131 flower-visits to yellow-flowered morphs, but only 20.9% of an observed total of 91 egg-layings were on yellow flowered morphs.

The extent to which the flower colour morph rations of the progeny of the Beenham, Singleton and Sketty B populations differed from those that would be produced by random outcrossing, calculated from Hardy-Weinberg expected frequencies, was investigated by growing seedlings of the flowering stage from large population samples of seed collected during 1975 (Table 4). The progeny of yellow-flowered morphs in the Beenham population showed a small but probably not significant excess of yellows. Thus the relatively small amount of preferential pollination by *Pieris rapae* and *Eristalis tenax* that was observed at Beenham did not significantly affect the morph ratio in the sample. As the yellow-flowered morphs are homozygous recessives, the Beenham progeny analysis also confirms that there was no significant self-compatibility in the population.

Table 4. Numbers of white- and yellow-flowered plants among seedlings grown from spontaneous seed produced in polymorphic populations of *R. raphanistrum* during 1975

| Parental population | Flower-colour of progeny | | | |
| | White | | Yellow | |
	Observed	Expected	Observed	Expected
Beenham (7.3% Y)				
Yellow-flowered	106	112	46	40
	(no significant difference, $\chi^2 = 1.404$)			
Singleton (27.4% Y)				
Whole population	252	255	99	96
	(no significant difference, $\chi^2 = 0.11$)			
Sketty B (60.8% Y)				
Yellow-flowered	15	24	94	85
	(excess of yellows, $P<0.05$, $\chi^2 = 4.34$)			
White-flowered	98	87	34	45
	(excess of whites, $P<0.05$, $\chi^2 = 4.15$)			
Whole population	47	59	105	92
	(excess of yellows, $P<0.05$, $\chi^2 = 4.50$)			

The Singleton progeny analysis suggests that the observed preferential pollination of white-flowered morphs by *Bombus pascuorum* may have balanced the observed preferential pollination of yellow-flowered morphs by *Pieris* spp., *Apis mellifera* and *Eristalis arbustorum* (Table 3); unfortunately, separate progeny analyses were not performed for the yellow-flowered and white-flowered morphs in the Singleton population. At Sketty B, however, such analyses showed significant excesses of both whites from white and yellows from yellow, with an overall excess of yellow, suggesting that the influence of yellow-seeking preferential pollinators may have predominated, although alternative explanations cannot be excluded.

In any case it is probable that the discrimination between yellow-flowered and white-flowered morphs that is shown by many of the insect pollinators of *R. raphanistrum* plays an important role both in maintaining its flower colour polymorphism, and in determining the frequency of the morphs. It seems possible that the recessive yellow-flowered morph could be selectively disadvantageous, except at low frequencies, in regions where *Sinapis arvensis*, *S. alba* and similar insect purple crucifers with relatively accessible nectar are abundant. Long-tongued bumble-bees such as *Bombus pascuorum* and *B. lapidarius* might preferentially major on white-flowered morphs of *Raphanus raphanistrum* in these areas, because the nectar of *R. raphanistrum* is less accessible to short-tongued visitors than that of *Sinapis arvensis*. Yellow-flowered *R. raphanistrum* might, however, be confused with

Sinapis spp., and it would be advantageous to the long-tongued *Bombus* spp. to major on the easily distinguishable white-flowered morph of *R. raphanistrum*.

The yellow-flowered morph could be maintained at a low frequency in these populations by the frequency-dependent effect (e), with an increasing relative frequency of visits by *Pieris* spp., *Eristalis* spp. or other preferential pollinators as its abundance in the population falls. Mixed populations (with fairly high frequencies of both morphs) might be maintained not as a balanced polymorphism but as a fluctuating and/or mosaic polymorphism, (f) and/or (g), with seed storage in the soil helping to maintain the relative stability of the mixture (Epling *et al.*, 1960).

The field observations provide some support for these hypotheses, although the decrease in preference for the yellow-flowered morph with decreasing frequency in the population apparently shown by *Pieris* spp., indicates that there is a particular need for work on the behaviour of pollinators in populations with low frequencies of the yellow-flowered morph. The interrelationships of the flower colour and flower veining polymorphisms, and the factors maintaining the flower veining polymorphism, are also problematic.

REFERENCES

BAKER, H. G., 1948. Stages in invasion and replacement demonstrated by species of *Melandrium*. *Journal of Ecology, 36:* 96–119.

BATEMAN, A. J., 1951. The taxonomic discrimination of bees. *Heredity, 5:* 271–278.

BISHOP, L. G. & CHUNG, D. W., 1972. Convergence in visual sensory capabilities in a pair of Batesian mimics. *Journal of Insect Physiology, 18:* 1501–1508.

CLEMENT, W. M., 1965. Flower color, a factor in attractiveness of alfalfa clones for honeybees. *Crop Science, 5:* 267–268.

DAUMER, K., 1956. Reizmetrische Untersuchungen des Farbensehens der Bienen. *Zeitschrift für vergleichende Physiologie, 38:* 413–478.

DAUMER, K., 1958. Blumenfarben, wie sie die Bienen sehen. *Zeitschrift für vergleichende Physiologie, 41:* 49–110.

DRONAMRAJU, K. R., 1960. Selective visits of butterflies to flowers; a possible factor in sympatric speciation. *Nature, 186:* 178.

EPLING, C. & DOBZHANSKY, T., 1942. Genetics of natural populations. VI. Microgeographic races in *Linanthus parryae. Genetics, 27:* 317–332.

EPLING, C., LEWIS, H. & BALL, F. M., 1960. The breeding group and seed storage; a study in population dynamics. *Evolution, 14:* 238–255.

ERICKSON, E. H., 1975a. Honeybees and soybeans. *American Bee Journal, 115:* 351–353.

ERICKSON, E. H., 1975b. Variability of floral characteristics influences honeybee visitation to soybean blossoms. *Crop Science, 15:* 767–771.

FAULKNER, G. J., 1976. Honeybee behaviour as affected by plant height and flower colour in Brussels sprouts. *Journal of Apiculture Research, 15:* 15–18.

FREE, J. B. & WILLIAMS, I. H., 1973. The foraging behaviour of honeybees (*Apis mellifera*) on Brussels sprouts (*Brassica oleracea* L.). *Journal of Applied Ecology, 10:* 489–499.

GOPLEN, B. P. & BRANDT, S. A., 1975. Alfalfa flower colour associated with differential seed set by leafcutter bees. *Agronomy Journal, 67:* 804–806.

HARBERD, D. J. & KAY, Q. O. N., 1975. *Raphanus*. In C. A. Stace (Ed.), *Hybridization and the Flora of the British Isles*. London: Academic Press.

HEINRICH, B., 1976. The foraging specialisations of individual bumblebees. *Ecological Monographs, 46:* 105–128.

HOROVITZ, A. & COHEN, Y., 1972. Ultraviolet reflectance characteristics in flowers of crucifers. *American Journal of Botany, 59:* 709–713.

JOHNSTON, G., 1829. *A Flora of Berwick upon Tweed*. Edinburgh.

KAY, Q. O. N., 1976. Preferential pollination of yellow-flowered morphs of *Raphanus raphanistrum* by *Pieris* and *Eristalis* spp. *Nature, 261:* 230–232.

KAUFFELD, N. M. & SORENSEN, E. L., 1971. Interrelations of honeybee preference of alfalfa clones and flower color, aroma, nectar volume and sugar concentration. *Kansas Agricultural and Experimental Station. Research Publication*, No. 163: 1–14, Manhattan, Kansas.

KENDALL, D. A. & SOLOMON, M. E., 1973. Quantities of pollen on the bodies of insects visiting apple blossom. *Journal of Applied Ecology, 10;* 627–634.

KUGLER, H., 1955. *Einführung in der Blütenökologie.* Stuttgart: Fischer.

LELEJI, O. I., 1973. Apparent preference by bees for different flower colours in cowpeas (*Vigna sinensis* (L.) Savi ex Hassk.), *Euphytica, 22:* 150–153.

LEVIN, D. A., 1969. The effect of corolla color and outline on interspecific pollen flow in *Phlox. Evolution, 23:* 444–445.

LEVIN, D. A., 1972a. The adaptedness of corolla-color variants in experimental and natural populations of *Phlox drummondii. American Naturalist, 106:* 57–70.

LEVIN, D. A., 1972b. Low frequency disadvantage in the exploitation of pollinators by color variants in *Phlox. American Naturalist, 106:* 453–460.

LEVIN, D. A. & BERUBE, D. E., 1972. *Phlox* and *Colias:* the efficiency of a pollinating system. *Evolution, 26:* 242–250.

LEVIN, D. A. & KERSTER, H. W., 1970. Phenotypic dimorphism and population fitness in *Phlox. Evolution, 24:* 128–134.

LEVIN, D. A. & KERSTER, H. W., 1973. Assortative pollination for stature in *Lythrum salicaria. Evolution, 27:* 144–152.

LEVIN, D. A. & SCHAAL, B. A., 1970. Corolla color as an inhibitor of interspecific hybridization in *Phlox. American Naturalist, 104:* 273–283.

LLOYD, D. G., 1969. Petal color polymorphism in *Leavenworthia* (Cruciferae). *Contributions from the Gray Herbarium of Harvard University, 198:* 9–40.

MAZOKHIN-PORSHNYAKOV, G. A., 1962. Colorimetric index of trichromic bees. *Biofizika, 7:* 211–217.

MOGFORD, D. J., 1974a. Flower colour polymorphism in *Cirsium palustre.* 1. *Heredity, 33:* 241–256.

MOGFORD, D. J., 1974b. Flower colour polymorphism in *Cirsium palustre.* 2. Pollination. *Heredity, 33:* 257–263.

PEDERSEN, M. W., 1967. Cross pollination studies involving three purple-flowered alfalfa, one white-flowered line, and two pollinator species. *Crop Science, 7:* 59–62.

PEDERSEN, M. W. & TODD, F. E., 1949. Selection and tripping in alfalfa clones by nectar collecting honey bees. *Agronomy Journal, 41:* 247–249.

PERRING, F. H. (Ed.), 1958. *Critical Supplement to the Atlas of the British Flora.* London: Nelson.

PERRING, F. H. & WALTERS, S. M. (Eds), 1962. *Atlas of the British Flora.* London: Nelson.

RAY, J., 1724. *Synopsis Methodica Stirpium Britannicarum,* 3rd ed. London.

SAMPSON, D. R., 1964. A one-locus self-incompatibility system in *Raphanus raphanistrum. Canadian Journal of Genetics and Cytology, 6:* 434–445.

SAMPSON, D. R., 1967. Frequency and distribution of self-incompatibility alleles in *Raphanus raphanistrum. Genetics, 56:* 241–251.

SMITH, J. E., 1804. *English Botany,* Vol. 12. London.

STRUWE, G., 1972a. Spectral sensitivity of the compound eye in butterflies (*Heliconius*). *Journal of Comparative Physiology, 79:* 191–196.

STRUWE, G., 1972b. Spectral sensitivity of single photoreceptors in the compound eye of a tropical butterfly (*Heliconius numata*). *Journal of Comparative Physiology, 79:* 197–201.

WALLACE, A. R., 1889. *Darwinism,* 2nd ed. London.

Pollination and flower colour polymorphism, with special reference to *Cirsium palustre*

D. J. MOGFORD

Botany Department, Rhodes University, Grahamstown, South Africa

In seacliff and mountain populations, the thistle *Cirsium palustre* is polymorphic for flower colour, occurring as a white flowered morph in addition to the usual purple flowered type. Since the white morphs are subject to preferential pollination, the occurrence of the polymorphism is interpreted as an adaptive response to conditions of low pollination.

KEY WORDS:—*Cirsium*—South Wales—flower colour polymorphism—*Bombus*—preferential pollination—cline.

CONTENTS

INTRODUCTION

Comparatively few types of polymorphism suggest to us the reasons for their existence. In flower colour polymorphisms, however, the general association between flower colour and insect pollination suggests certain selective factors which may be involved.

By restricting the discussion to polymorphism in the sense of Ford (1940), we exclude the occurrence of variants maintained merely by recurrent mutation. We may note, however, that pollinator discrimination against mutant colour types might be one of the factors which prevent their establishment. It may be relevant in this connection that white flowered variants have been recorded as having failed to set seed in both *Clarkia cylindrica* (Lewis, 1953) and *Pseudomuscari azureum* (Garbari, 1972). An unusual instance was cited by Darwin (1876), who recorded that the flowers of a white flowered variant of *Aconitum napellus* were perforated by bees (nectar-robbing), whereas those of the typical blue-flowered form were not. Darwin considered that this was likely to be caused by the blue variety being distasteful, and pointed out that, in view of the protandry of the species its effect would be to reduce the seed set of the white flowered type.

The possible roles of pollination in the maintenance of flower colour polymorphism are several, but we may begin with the presumption that the colour types may differ in their attractiveness to pollinating insects. Discrimination of this type has been recorded in *Lantana camara* (Dronamraju, 1960), *Leavenworthia crassa* (Lloyd, 1969) and *Raphanus raphanistrum* (Kay, 1976). Records are also available of colour discrimination in connection with the changes in flower colour that sometimes accompany floral development. Thus, in *Pulmonaria officinalis*, in which the flowers are at first red but later blue, Müller (1883b) observed that a bee species confined its visits mainly to the red stage. Similarly, F. Müller observed that in a Brazilian *Lantana* species, in which the flowers change over successive days from yellow to orange and then purple, certain butterflies visit the yellow and orange flowers, and others exclusively the yellow, but none the purple (Darwin, 1877).

One effect which follows from the differences in attractiveness of morph types is that the more attractive type may be visited first. Little evidence is available that this is of general importance. In the dioecious *Ribes alpinum*, however, in which flower colour differences between plants are associated with the sexual differences, the colour differences are said to encourage pollinators to visit the plants before the female—an effect produced in the protandrous *Lonicera periclymenum* by change in colour during floral development (Knuth, 1906).

Secondly, should a variant be of greater general attractiveness than the widespread type, the overall level of pollination of the population containing it will be increased, since higher numbers of foragers will be attracted and foraging will be encouraged for longer periods. Selection for populations homozygous for the more attractive type may be counterbalanced by, for example, some physiological disadvantage or by heterozygous advantage. In any case, a polymorphism of this type may be maintained in particular by conditions of limited pollination. Thus, Hocking (1968) observed that in a region of the Canadian Arctic in which pollinating insects are present in limited numbers, there occur white forms of *Saxifraga*, *Erysimum* and *Pedicularis* species. It is probably for this reason that *Cirsium palustre* is polymorphic for flower colour (see later). The flower colour polymorphism of *Phlox pilosa* may also be explained on this basis, though a form of prevention of interspecific pollination (competitive exclusion through 'colour displacement') is also of importance here (Levin & Kerster, 1967).

Thirdly, is the condition where pollinating species differ with respect to attraction to flower colours, so providing a degree of divergence in pollinating species between the morph types. The earlier pollination ecologists cited many examples of this. Müller (1881) recorded that at high altitude, the typically red flowered *Trifolium pratense* exhibits a white flowered variety, which he stated to be preferentially pollinated by Lepidoptera, which are of greater importance as pollinators at such altitude. In the same work, Müller noted that *Primula farinosa* is more brightly coloured in the Alps than in Pomerania, and attributed this to the species being adapted to butterflies in the Alps, but to bees in Pomerania. Müller later stated (1883a) that alpine *Primula* species possess a general tendency to be red flowered as an adaptation to butterfly pollination.

Knuth (1906) noted that *Gymnadenia* (*Habenaria*) *conopsea* is generally purple flowered but is sometimes white, and considered these types to indicate differential adaptation to butterfly and moth pollination respectively. The same explanation has been employed by Müller (1878) for his observation that alpine populations in both *Daphne striata* and *Anacamptis pyramidalis* exhibit intrapopulational variation in colour between rose-red and white. Müller observed that *Crocus vernus* and *Gymnadenia odoratissima* vary similarly, but only between pale rose and white, which he considered as representing a greater range of adaptation to moths.

Such specialisation of pollinating species indicates the possibility that a reproductive isolation between the morph types may become established. Thus, a polymorphism of this type may be suited as a medium for sympatric evolutionary divergence. Such effects may also be provided without any such specialisation if the colour differences between the morphs are sufficient to elicit search image responses for particular colour types by the pollinators. These effects will be related to the magnitude of the colour differences involved—any initial differences effecting partial separation may later be improved upon. Certainly, in the case of *Anagallis arvensis*, one is struck by the contrast between the subspecies *arvensis*, which in Britain is typically red flowered, and the blue flowered subspecies *foemina*, and I have observed that where the two types were growing together, they were almost completely isolated from each other by dipteran pollinators. Such isolation could well help to maintain the vegetative differences between these subspecies, though F1 sterility is also of importance in this particular case (Marsden-Jones & Weiss, 1959).

THE FLOWER COLOUR POLYMORPHISM OF *CIRSIUM PALUSTRE*

Cirsium palustre is a biennial, sexually reproducing thistle which occurs native in Europe and west Asia, and as an introduced species elsewhere. It grows in populations of a few hundred to occasionally a few thousand individuals, whose borders are usually well defined due to ecological limitations, and which can therefore generally be clearly demarcated for the purpose of population scoring. While the species possesses a small degree of self-compatibility, in most natural populations the species is vigorously crosspollinated by bumblebees and the seed set is high.

The polymorphism for flower colour is such that while populations in almost all cases show a majority of the typical purple flowered type, they may in addition show a white flowered type, and, at respectively lower frequencies, approximately two categories of partially coloured plants which are referred to as "intermediate" and "purplish white". The genetics of the polymorphism is complex; while the white morphs are exclusively homozygous, the intermediate and purple morphs are each either homozygous or heterozygous (Mogford, 1974a).

The distribution of the polymorphism in England and Wales has been studied at length. Throughout inland, lowland sites, the species is uniformly almost completely purple flowered. Seacliff populations, however, are highly polymorphic. The situation with regard to altitude is also remarkable; on travelling westwards from Oxford, across the Cotswolds and onto the Welsh foothills one encounters only monomorphic purple populations, a situation which continues to a height of 280 m in the Welsh mountains. On travelling just 20 m higher one finds a striking change; virtually all populations contain high frequencies of the white morph. There is, therefore, a "barrier effect"; a barrier which, though associated with a mere 20 m of altitude, holds true throughout southern and central Wales, and, with minor modification, in north Wales as well (Fig. 1).

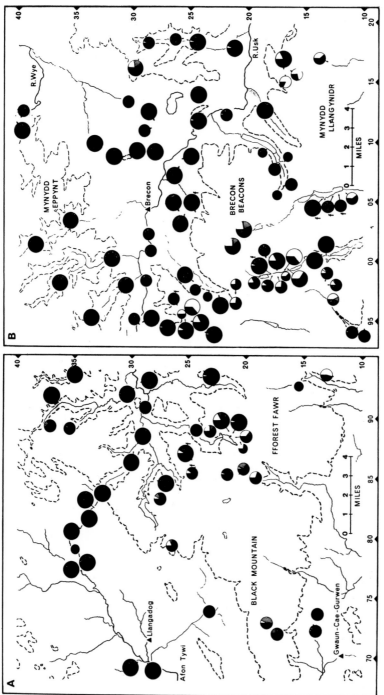

Figure 1. Distribution of the polymorphism in southern mid-Wales. Populations are represented by circles, related in size to population size and divided into sectors proportional to the frequency of the different morph types.

Population size	Circle Size	Morph type	Sector
150 or less	small	white	white
151–500	medium	purplish white	lightly stippled
more than 500	large	intermediate	heavily stippled
		purple	black

Dotted lines represent 1000-ft contour. Forest populations denoted by 'f'.

Firstly it is necessary to indicate the selective factors other than pollination to which we know the polymorphism is subject (Mogford, 1972, 1974a).

Comparisons of morph frequency between subpopulations differing in exposure has indicated that within both seacliff and mountain populations intrapopulational, selection occurs such that the frequency of white morphs is least in the most exposed segments. The influence of exposure is especially apparent among the populations of the Gower Peninsula in South Wales, in which the distribution of white morphs is clinal, their frequency being lowest in those populations which are the most exposed by virtue of greater cliff height and proximity to the tip of the peninsula.

Comparisons of the morph frequency and population size of each of 45 Gower seacliff populations over four consecutive seasons has indicated that reduction in population size is accompanied by increase in the frequency of the purple morphs, and a corresponding decrease in that of each of the other types. Increase in population size is accompanied by the reverse effects.

One likely explanation of this is that improvement in environmental conditions allows an increase in population size and a relaxation in selection against the non-purple morphs, whereas deterioration in environmental conditions produces the opposite effects. This type of influence has been proposed previously by Ford & Ford (1930) following their work on variability in the butterfly *Melitaea aurinia*.

While these factors are applicable to particular instances, they fail to account for the overall distribution of the polymorphism. Seacliffs and mountains are the most exposed habitats in which the species occurs; yet it is in precisely these habitats that the flower colour is polymorphic.

Both mountains and seacliffs possess climatic conditions likely to reduce the degree of insect pollination; both are subject to high exposure, while mountains are in addition subject to low temperatures and large amounts of mist and rain, these various conditions being likely to diminish both visitor activity and the total visitor population (Mogford, 1974b).

It is significant that the altitude on mountains at which the white morphs suddenly appear at high frequency, 300 m, coincides with the general limit of ground enclosure by farmers. This limit bears testimony to the sudden deterioration in climatic conditions at this altitude, the cloud base at this height resulting in a marked increase of mist and rain. Pollination of *C. palustre* is almost exclusively due to bumblebees, and since the activity of bumblebees is quickly stopped by rain and heavy mist, the altitude at which white morphs increase in frequency is that at which their pollination may be assumed to be most suddenly limited.

In conditions of limited pollination, self-compatible, or unduly attractive morphs may be favoured. Greenhouse experiments indicate that colour morphs do not differ in self compatibility, so experiments were designed to determine whether morphs differed in their attractiveness to visitors.

These experiments were performed on six populations, five on mountains and the other (Rhossili Down) on a seacliff. The visits of bees within each area were followed, the bees being classified by species and in some cases sex, and for each such category a value obtained for the ratio of morph types visited. The significance of the degree of discrimination displayed by each bee type was then calculated for each morph type in each population. This was performed using chi-squared tests

D. J. MOGFORD

Table 1. Preferential pollination

	Morph and visits ratios					Discrimination probability				
	P	I	ppW	W	n	P	I	ppW	W	f %
(a) Mynydd Llangattwg (SO 159166)										
Morph ratio	66.5	0.4	32.7	0.4	245					
B. agrorum	67.5	0.0	32.3	0.1	671	N.S.	—	N.S.	—	79.1
B. lapidarius (♂)	63.6	0.0	36.4	0.0	580	N.S.	—	N.S.	—	5.9
B. lapidarius (♀)	75.4	0.0	24.6	0.0	427	**	—	(**)	—	7.6
B. lucorum	53.5	0.0	46.5	0.0	636	(***)	—	***	—	6.6
(b) Mynydd Troed (SO 171295)										
Morph ratio	61.3	16.2	21.7	0.7	678					
B. agrorum	58.0	18.7	22.8	0.5	653	N.S.	N.S.	N.S.	—	44.0
B. lapidarius (♂)	65.2	16.7	18.2	0.0	402	N.S.	N.S.	N.S.	—	29.0
B. lapidarius (♀)	56.8	20.0	23.2	0.0	375	N.S.	N.S.	N.S.	—	9.4
B. lapponicus	50.9	17.9	31.1	0.0	636	(***)	N.S.	***	—	6.2
B. lucorum	56.5	19.5	24.0	0.0	563	(*)	*	N.S.	—	11.3
(c) Beacons Forest (SN 989180)										
Morph ratio	95.6	0.0	0.0	4.4	413					
B. pratorum (♀)	96.3	—	—	3.7	514	N.S.	—	—	N.S.	28.1
(d) Rhossili Down (SS 41 896)										
Morph ratio	76.7	0.0	0.8	21.4	271					
A. mellifera	71.4	—	2.4	26.2	764	N.S.	—	N.S.	N.S.	c. 100
(e) Cantref (SN 992168)										
Morph ratio	74.5	0.0	0.0	25.5	699					
B. agrorum	44.1	—	—	25.9	711	N.S.	—	—	N.S.	14. 1
B. lapponicus	71.7	—	—	28.3	669	N.S.	—	—	N.S.	1.3
B. lucorum	70.5	—	—	29.5	1147	(*)	—	—	*	79.6
B. pratorum (♀)	58.9	—	—	41.1	615	(***)	—	—	***	5.0
(f) Forest Lodge (SN 961247)										
Morph ratio	36.0	0.0	0.0	64.0	356					
B. agrorum (♂)	37.3	—	—	62.7	416	N.S.	—	—	N.S.	51.7
B. agrorum (♀)	37.6	—	—	62.4	431	N.S.	—	—	N.S.	23.0
B. lapidarius (♂)	26.3	—	—	73.7	646	(***)	—	—	***	8.2
B. lapidarius (♀)	18.5	—	—	81.5	433	(***)	—	—	***	1.1
B. lapponicus	11.6	—	—	88.4	362	(***)	—	—	***	1.0
B. lucorum	17.5	—	—	82.5	618	(***)	—	—	***	15.0

Morph and visit ratios listed as percentage values based on sample size n. Species frequency values, f, calculated from mean values on each day of observation.

Probability values calculated as described in text. Unbracketed probability values indicate morph types subject to preferential pollination, and bracketed values indicate morph types discriminated against.

*=significant at 5% level; **=significant at 1% level; ***=significant at 0.1% level.

to assess the difference between the observed and expected values for the ratio of the number of morphs of a particular type visited to the total number of alternative morphs present in each population (Table 1).

It was found that discriminatory visiting was a frequent phenomenon, and in almost all cases involved preferential visiting of the white morphs. Thus, of twelve cases in which visiting of the white morphs was studied, six showed preferential visiting of this type, and in no cases were they discriminated against. In contrast, of the 21 cases in which visiting of the purple morphs was studied, only one involved preferential visiting of this type, whereas in nine cases they were discriminated against.

The experimental method naturally involved certain variables. The degree of discrimination varied between species, between sexes, and between the same sex in different populations. This indicates that the net degree of discrimination will be affected by seasonal and daily fluctuations in the relative abundance of visitors. However, although the numbers of foragers were reduced dramatically by unfavourable conditions, the relative proportions of the various types present appeared unaffected.

It may be concluded that white morphs are subject to preferential pollination. This will produce assortative pollination among white morphs, effecting some inbreeding, augmenting that already imposed on the population limitations in pollination. However, white morphs will show relatively more cross-pollination than other morphs, which may be assumed to be selectively advantageous (for example, greenhouse experiments show that cross pollination greatly increases seed production (Mogford, 1974b)).

The general distribution of the polymorphism is consistent with a relative increase in the fitness of the white morph type as a consequence of preferential pollination in conditions of limited pollination.

DISCUSSION

The association of the flower colour polymorphism of *Cirsium palustre* with altitude is of particular interest since altitudinal variation in flower colour is a very general phenomenon. The types of flower colour variation involved, however, are various. Thus, whereas in *Cirsium palustre* white flowered plants increase in frequency relative to the purple type on mountains, the reverse effect is seen in the occurrence at high altitude of purplish forms in the typically white flowered *Euphrasia salisburgensis* (Tutin *et al.*, 1964), and of blue forms in an otherwise white flowered *Ageratum* species in Java (Weevers, 1952). Indeed, Weevers noted from a study of floras that whereas the proportion of different coloured species is relatively constant in most countries, in both Switzerland and Java the proportion of blue flowered species is higher among the plants of mountain regions than among the plants of these countries taken as a whole. This effect is seen also in the occurrence in *Myosotis alpestris* of darker blue flowers than those of lowland *Myosotis* species, and by the tendency within *M. palustris* for mountain plants to produce deeper blue flowers than those on lowlands (Knuth, 1906).

Altitudinal variation in flower colour occurs also in *Linaria reflexa*, which is typically yellow flowered, but is white on mountains (Engler & Prantl, 1887); *L. aeruginea*, which is usually yellow, but is always yellow at high altitude on the Sierra Nevada, Spain, (Tutin *et al.*, 1964); *Melampyrum pratense*, in mountain populations of which a form with crimson corolla tips occurs in addition to the usual yellow types (Smith, 1963); and *Encelia farinosa*, a typically yellow flowered composite in which a form with brownish-purple disc florets occurs in greater frequency at higher altitude (Khyos, 1971).

However, it is not yet possible to estimate the involvement of pollination in these various cases. In particular, the physiological influences of mountain conditions are also likely to be of importance. Thus, Garbari (1970) observed that Corsican populations of *Brimeura fastigiata* are pink flowered in lowland regions, but white flowered on mountains, and that the white flowered plants when grown in lowland conditions become pink flowered. Similarly, Baldwin (1970) recorded that among mountain populations of *Cypripedium acaule* in North America, in which white flowered plants are present in addition to the pink flowered type uniformly found elsewhere (Anderson, 1936), a surfeit of white flowered plants occurred following an exceptionally cold winter, and concluded that the production of white flowers was likely to have been a physiological effect of low temperature. It is notable also that *Campanula trachelium* is white flowered in certain regions of the Alps, but blue flowered in others (Kerner, 1894), for exposure of white flowered plants of the species to low temperature results in the production of blue flowers (Klebs, 1906).

Moreover, even in cases where pollination effects are involved in the maintenance of altitudinal variation in flower colour, these may result from different causes. The flower colour polymorphism of *Cirsium palustre*, apparently an adaptation to limited pollination, may be atypical, since the Welsh mountains are unusual in respect of the high amounts of mist and rain encountered there at relatively low altitudes. Other cases of altitudinal variation may represent adaptation towards differences in pollinating species at different altitudes.

There occurs another habitat where pollination effects are likely to be of importance in the maintenance of flower colour polymorphism, namely in woodlands. Anderson (1936) remarked that in New England forests, the proportion of white flowered plants in *Hepatica acutiloba* and *Dicentra cucullaria* is relatively higher than in the less dense Ozark woodlands, and attributed this to the paler flowers being more attractive to visitors in dull light. This is evidently possible in view of the observation of Hulkkonen (1928) that at low light intensity bees visit white flowers in preference to coloured ones. It is relevant also that moth pollinated flowers, which are mainly light pollinated, have long been known to be generally pale in colour (Sprengel, 1793).

In *Cirsium palustre* the opposite situation obtains, forest populations tending towards a monomorphic purple condition regardless of altitude (Fig. 1). One explanation, following Müller (1883a), might be that in woodlands visiting may continue even during unfavourable weather. It seems more probable, however, that in woodlands preferential pollination may be of less importance relative to physiological selective effects acting upon the polymorphism (Mogford, 1974b).

It is evident, therefore, that flower colour polymorphisms present a host of examples of variation, which have yet to be examined from the viewpoint of pollination ecology. Although an abundance of examples exists of situations where flower colour polymorphisms have been related to fixation effects, recombinational variance following hybridisation, and ecological factors which have no apparent bearing on pollination, it can nevertheless be anticipated that the further application of more exact tests for pollinator discrimination will yield valuable results in this field.

ACKNOWLEDGEMENT

I wish to thank the editor and publishers of *Heredity* for allowing me to reproduce Fig. 1 and Table 1.

REFERENCES

ANDERSON, E., 1936. Colour variation in eastern North American flowers as exemplified by *Hepatica acutiloba*. *Rhodora, 38:* 301–304.

BALDWIN, J. T., 1970. White phase in flower development in *Cypripedium acaule*. *Rhodora, 72:* 142–143.

DARWIN, C., 1876. *The Effects of Cross and Self Fertilisation in the Vegetable Kingdom*. London: Murray.

DARWIN, C., 1877. Fritz Müller on flowers and insects. *Nature, 17:* 78–79.

DRONAMRAJU, K. R., 1960. Selective visits of butterflies to flowers: a possible factor in sympatric speciation. *Nature, 186:* 178.

ENGLER, A. & PRANTL, K., 1887. *Die Natürlichen Pflanzenfamilien*. Leipzig

FORD, E. B., 1940. Polymorphism and taxonomy. In J. Huxley (Ed.), *The New Systematics*. Oxford: Clarendon Press.

FORD, H. D. & FORD, E. B., 1930. Fluctuation in numbers and its influence on variation in *Melitaea aurinia Transactions of the Royal Entomological Society of London, 78:* 345–351

GARBARI, F., 1970. Il genere *Brimeura* Salisb. (Liliaceae). *Memorie della Societa toscana di scienze naturali (Ser. B), 77:* 12–36.

GARBARI, F., 1972. Note sul genere "*Pseudomuscari*" (Liliaceae). *Webbia 27:* 369–381.

HOCKING, B., 1968. Insect-flower associations in the high arctic with special reference to nectar. *Oikos 19:* 359–388.

HULKKONEN, O., 1928. Zur biologie der südfinnischen Hummeln. *Annales Universitatis fennicae aboensis (Ser. A), 3:* 1–81.

KAY, Q. O. N., 1976. Preferential pollination of yellow-flowered morphs of *Raphanus raphanistrum* by *Pieris* and *Eristalis* species. *Nature, 261:* 230–232.

KERNER VON MARILAUN, A., 1894. *The Natural History of Plants*. London.

KLEBS, G., 1906. Über variationen der blüten. *Fahrbuch für wisenschaftliche Botanik, 42:* 155–320.

KNUTH, P., 1906. *Handbook of Flower Pollination*. Oxford.

KYHOS, D. W., 1971. Evidence of different adaptations of flower colour variants of *Encelia farinosa*. *Madroño, 21:* 49–61.

LEVIN, D. A. & KERSTER, H. W., 1967. Natural selection for reproduction isolation in *Phlox*. *Evolution, 21:* 679–687.

LEWIS, H., 1953. The mechanism of evolution in the genus *Clarkia*. *Evolution, 7:* 1–20.

LLOYD, D. G., 1969. Petal colour polymorphism in *Leavenworthia* (Cruciferae). *Contributions from the Gray Herbarium of Harvard University, 198*.

MARSDEN-JONES, E. M. & WEISS, F. E., 1959. The genetics and pollination of *Anagallis arvensis* subsp. *arvensis* and *Anagallis arvensis* subsp. *foemina*. *Proceedings of the Linnean Society of London, 171:* 27–29.

MOGFORD, D. J., 1972. *The ecological genetics of flower colour variation in* Cirsium palustre. D.Phil. thesis, University of Oxford.

MOGFORD, D. J., 1974a. Flower colour polymorphism in *Cirsium palustre*. 1. *Heredity, 33:* 241–256.

MOGFORD, D. J., 1974b. Flower colour poymorphism in *Cirsium palustre*. 2. Pollination. *Heredity, 33:* 257–263.

MÜLLER, H., 1878. Die Insekten als unbewusste Blumenzüchter. *Kosmos, 3:* 403–426.

MÜLLER, H., 1881. *Die Alpenblumen, ihre Befruchtung durch Insekten und ihre anpassungen an dieselben*. Leipzig.

MÜLLER, H., 1883a. *The Fertilisation of Flowers*. London.

MÜLLER, H., 1883b. The effect of the change of colour in the flowers of *Pulmonaria officinalis* upon its fertilisers. *Nature, 28:* 81.

SMITH, A. J. E., 1963. Variation in *Melampyrum pratense* L. *Watsonia, 5:* 336–367.

SPRENGEL, C. K., 1793. *Das entdeckte Geheimnis der Natur im Bau und in der Befruchtung der Blumen*. Berlin.

TUTIN, T. G., *et al.* (Eds), 1964. *Flora Europaea*. Cambridge University Press.

WEEVERS, T., 1952. Flower colours and their frequency. *Acta Botanica Neerlandica, 1:* 81–92.

The effect of insect behaviour on hybrid seed production of Brussels sprouts

G. J. FAULKNER

National Vegetable Research Station, Wellesbourne, Warwick, U.K.

F_1 hybrid Brussels sprouts give greater plant uniformity, improved sprout quality and higher yields of marketable sprouts than do non-hybrid cultivars. F_1 hybrids are now widely grown and a considerable number of cultivars are available. However, there are several problems concerned with large-scale seed production of F_1 hybrids.

F_1 hybrid seed is produced by growing together two different parent inbred lines, relying mainly upon honeybees to effect cross-pollination. The inbred lines must be cross-compatible and preferably both self- and sib-(sister–brother) incompatible. Seed resulting from self- or sib-pollinations produces plants of poor quality and reduced sprout yield; these are collectively known as "sibs". Seed of almost all F_1 hybrid cultivars of Brussels sprouts contains sibs, the number varying not only from cultivar to cultivar but also between seed lots of the same cultivar, according to site and year of production.

There appear to be two main factors that affect the amount of sib seed produced; the variation in the self-incompatibility of the inbred lines, and the pollinating behaviour of insects: honeybees and blowflies have been studied. The only solution to the problem of unstable self-incompatibility is to breed new inbred lines which have stronger and more stable self-incompatibility. However, this is a fairly long-term process which is the more difficult since breeders have to bear in mind that the end product must be agronomically satisfactory.

Honeybees contribute to the sib problem because, as shown by work at the National Vegetable Research Station, they are highly selective between the plants of each inbred line. Collectively, they do not appear to prefer one line or the other, but individuals remain constant to one or the other inbred of their choice. In many different pollination experiments, results have shown that selfing movements are far in excess of crossing movements; a self : cross ratio of as much as 30 : 1 has been calculated. Honeybees are able to distinguish differences in plant characteristics between lines, and, in particular, differences in flower colour and plant height. Other factors, such as differential flowering of the inbreds, and winter losses can affect the amount of pollen available for distribution by honeybees.

Blowflies, on the other hand, are random in their behaviour, and have no preference for either line. Experiments have shown that this behaviour results in more efficient cross-pollination and that fewer sibs are produced. Blowflies have to be confined to the plot, whereas honeybees may fly freely.

Work has recently been carried out at the National Vegetable Research Station into a comparison of the effect of honeybees and blowflies on seed yield, percentage sibs and the cost of seed production under different pollination structures. Compared with bees, blowflies drastically reduced the percentage of sibs from both parents under glass and under polythene tunnels, and to a lesser extent under nets and in the open. Fly-pollinated plants in polythene tunnels gave the lowest percentage sibs, the highest seed yield and the lowest cost of production of seed.

Index